English in Focus: Agriculture

ENGLISH IN FOCUS

Series Editors: J. P. B. ALLEN and H. G. WIDDOWSON

*Physical Science** J. P. B. Allen and H. G. Widdowson
*Mechanical Engineering** Eric H. Glendinning
*Workshop Practice** Alan Mountford
*Basic Medical Science** Joan Maclean
Agriculture Alan Mountford
Education Elizabeth Laird
Social Studies J. P. B. Allen and H. G. Widdowson
Biological Science Ian Pearson
Electrical Engineering and Electronics Eric H. Glendinning

* Advisory Editor: RONALD MACKIN

ENGLISH IN FOCUS

English in Agriculture

ALAN MOUNTFORD

OXFORD UNIVERSITY PRESS

Oxford University Press, Walton Street, Oxford OX2 6DP

OXFORD LONDON GLASGOW
NEW YORK TORONTO MELBOURNE WELLINGTON
KUALA LUMPUR SINGAPORE HONG KONG TOKYO
DELHI BOMBAY CALCUTTA MADRAS KARACHI
NAIROBI DAR ES SALAAM CAPE TOWN

ISBN 0 19 437514 5 (Student's Book)
ISBN 0 19 437506 4 (Teacher's Edition)

First published 1977
Reprinted 1979

Printed and bound in Great Britain by
Morrison & Gibb Ltd, London and Edinburgh

Contents

ix

Introduction xi

xviii

Unit 1 The Parts of a Plant and their Functions

I: READING AND COMPREHENSION 1
SOLUTIONS TO COMPREHENSION EXERCISE 2
EXERCISE A: Contextual reference 3
EXERCISE B: Rephrasing 4
EXERCISE C: Relationships between statements: consequence 4
II: LANGUAGE IN USE 5
EXERCISE A: Labelling a diagram 5
EXERCISE B: The definition of parts of a plant 6
EXERCISE C: General statements of function 7
III: GRAMMAR 7
EXERCISE A: The forms of definitions 7
EXERCISE B: The impersonal passive 8
IV: SUMMARY CHECK 11

Unit 2 The Life Cycle of a Plant

I: READING AND COMPREHENSION 12
SOLUTIONS TO COMPREHENSION EXERCISE 13
EXERCISE A: Contextual reference 14
EXERCISE B: Rephrasing 15
EXERCISE C: Relationships between statements: contrast 15
II: LANGUAGE IN USE 16
EXERCISE A: Definitions of processes 16
EXERCISE B: General statements of process 17

EXERCISE C: Statements of function and process 17
EXERCISE D: Definitions and descriptions of processes 18
III: GRAMMAR 24
EXERCISE A: Time expressions 24
EXERCISE B: Expressions of degree 26
IV: SUMMARY CHECK 27

Unit 3 The Origin and Composition of Soil 28
I: READING AND COMPREHENSION 28
SOLUTIONS TO COMPREHENSION EXERCISE 29
EXERCISE A: Contextual reference 30
EXERCISE B: Rephrasing 31
EXERCISE C: Relationships between statements: exemplification and explanation 32
II: LANGUAGE IN USE 33
EXERCISE A: Making tables from descriptions 33
EXERCISE B: Writing descriptions from tables 34
III: GRAMMAR 36
EXERCISE A: Comparative sentences 36
EXERCISE B: Contrastive sentences 39
EXERCISE C: Making comparisons by inference 39
IV: SUMMARY CHECK 41

Unit 4 Drainage and Irrigation 43
I: READING AND COMPREHENSION 43
SOLUTIONS TO COMPREHENSION EXERCISE 44
EXERCISE A: Contextual reference 46
EXERCISE B: Rephrasing 46
EXERCISE C: Relationships between statements: reinforcement and similarity 47
II: LANGUAGE IN USE 48
EXERCISE A: Classification and definition 48
EXERCISE B: Definition, description and identification 50
EXERCISE C: Classification in diagrams and paragraphs 51
EXERCISE D: Classification according to defining characteristics 53
III: GRAMMAR 55
EXERCISE: *to-infinitive* for the expression of purpose 55
IV: SUMMARY CHECK 57

Unit 5 Manures and Fertilizers 58
I: READING AND COMPREHENSION 58
SOLUTIONS TO COMPREHENSION EXERCISE 59
EXERCISE A: Contexual reference 61

EXERCISE B: Rephrasing 61
EXERCISE C: Relationships between statements: review 61
II: LANGUAGE IN USE 63
EXERCISE A: Conclusions based on observations 63
EXERCISE B: Generalizations 64
EXERCISE C: Recommendations 65
EXERCISE D: Predictions 66
III: GRAMMAR 68
EXERCISE A: Defining and non-defining relative clauses 68
EXERCISE B: Short-form relative clauses 70
IV: SUMMARY CHECK 72

Unit 6 The Control of Weeds and Plant Diseases 74
I: READING AND COMPREHENSION 74
SOLUTIONS TO COMPREHENSION EXERCISE 75
EXERCISE A: Contextual reference 76
EXERCISE B: Rephrasing 77
EXERCISE C: Relationships between statements 77
II: LANGUAGE IN USE 78
EXERCISE A: The identification and description of diseases 78
EXERCISE B: Recommendations 80
III: GRAMMAR 83
EXERCISE A: Noun + noun constructions 83
EXERCISE B: Participle + noun constructions 84
EXERCISE C: Complex noun phrases 85
IV: SUMMARY CHECK 86

Unit 7 Market Gardening 88
I: READING AND COMPREHENSION 88
EXERCISE A: Re-ordering paragraphs 88
EXERCISE B: Adding statements 89
EXERCISE C: Relationships between statements 89
EXERCISE D: Contextual reference 90
II: LANGUAGE IN USE 90
EXERCISE A: Directions and descriptions 90
EXERCISE B: Directions, descriptions and recommendations 92
EXERCISE C: Drawing conclusions 94
III: GRAMMAR 96
EXERCISE: Modal verbs 96
IV: SUMMARIZING 98

Unit 8 Animal Husbandry 100
PASSAGES FOR COMPREHENSION 100
PASSAGE I: *The Digestive System of Farm Animals* 100

PASSAGE II: *Food Requirements* 103
PASSAGE III: *Feedingstuffs* 106
PASSAGE IV: *Reproduction* 107
PASSAGE V: *Breeding* 110

Introduction

The aim of this book is to develop a basic knowledge of how English is used for communication in dealing with topics in agriculture. It is intended for students who already know how to handle the common English sentence patterns but who need to learn how these patterns are used to convey information and to conduct coherent discussion.

The exercises direct the student's attention to certain features of English which are commonly used in textbooks about agriculture. The aim is to provide the student with a strategy for reading more difficult texts in this subject area and to prepare him for making effective use of English in his own writing.

Although the emphasis is on English as a medium of expression for communicating ideas about agriculture, the basic elements of the language have not been neglected. Pattern practice is provided, particularly in the Language in Use and Grammar sections of each unit, but this kind of work is always presented in relation to a communicative context and not simply as an exercise in making sentences for their own sake.

This book does not aim at teaching the subject-matter of agriculture, and it does not aim at teaching grammatical structures and vocabulary as such. Its purpose is to show how language is used as a medium for the study of agriculture, and so to give students a grounding in one particular set of communication skills in English.

1 The Parts of a Plant and their Functions

I READING AND COMPREHENSION

[1]A plant is a living organism. [2]It is made up of different parts, each of which has a particular purpose, or specialized function. [3]If one part of the plant is not functioning properly the whole plant will suffer. [4]But we may cut flowers off the plant or prune the roots. [5]Such damage is only temporary and so the plant will continue to grow.

Study the following statements carefully and write down whether they are true or not true according to the information expressed above. Then check your answers by referring to the solutions at the end of the passage.*

(a) Different parts of a plant have specialized functions.
(b) Not all parts of a plant need function properly.
(c) If we cut flowers off a plant, the whole plant will suffer.

[6]The basic parts of a plant are the root system, which is below the ground, and the shoot system above. [7]The root of a plant has two main functions. [8]It takes in, or absorbs, water and minerals from the soil through the root hairs, which are single cells near the tip of each root. [9]The other main function of the root is to hold, or anchor, the plant firmly in position in the soil.

(d) The shoot system of the plant is below the ground.
(e) Water and minerals are absorbed through the root hairs.
(f) The plant is anchored in the soil by the root system.

[10]Plants such as sugar beet and carrots are able to store food in their roots. [11]In this way they can keep growing for more than one season. [12]In addition, plants such as clover and lucerne, known as 'legumes', have special bacteria which live on the roots. [13]These simple forms of life take nitrogen out of the air which is in the soil. [14]Such leguminous plants are usually ploughed under the soil. [15]By doing this the soil is made more fertile.

(g) Carrots can keep growing for more than one season.
(h) Special bacteria live on the roots of all plants.
(i) The soil is made fertile by ploughing clover and lucerne under the soil.

[16]The shoot system above the ground consists of the stem, the leaves, flowers and fruit. [17]One of the functions of the stem is to support the plant. [18]Another important function is to enable water and minerals to pass up from the roots to the leaves and flowers. [19]Organic materials such as sugar travel down the stem to the roots. [20]The leaves grow out of the side of the stem. [21]Their main job is to make food for the plant by the process known as photosynthesis. [22]For this process sunlight is necessary. [23]Water from the soil and carbon dioxide from the air are converted into sugars and other carbohydrates. [24]During the process oxygen is formed and released into the air. [25]The flower contains the reproductive organs of the plant. [26]The stamens produce the male sex cells, or spermatia, [29] which are carried in the pollen grains. [27]The carpel produces the female sex cells, or ovules. [28]The fruit, the ripened ovary of the flower, encloses the seeds and protects them while they are developing. [29] The seed itself consists of an embryo and foodstore. [30]The embryo is the part which will develop into another plant and the foodstore is necessary to provide nourishment for the young plant while it is growing.

Solutions

Add words or phrases from the text to complete the argument which shows whether the comprehension check is TRUE OR NOT TRUE. Note that a dotted line . . . . . . requires a phrase to be added, and a straight line ______ requires a word to be added. Numbers refer to the sentences in the text. The first one has been done as an example.

(a) Each of the different parts of a plant has a particular purpose. (2)
Each of the different parts of a plant has a particular purpose or specialized function. (2)

$\therefore$ a particular purpose $=$ a specialized function.

$\therefore$ Each of the different parts of a plant has a specialized function.

$=$ *Different parts of a plant have specialized functions.*

(b) The whole plant will suffer if ONE part is not . . . . (3)
i.e. If ALL . . . . . . are functioning properly the whole plant will ______ suffer.
If all parts of the plant are functioning properly the whole plant will not suffer.
i.e. All parts of the plant DO NEED to function properly.

* The following symbols are used in the solutions:

i.e. that is to say
e.g. for example
$=$ equals/means the same as
$\neq$ does not equal/mean the same as
$\therefore$ therefore

(c) We may ...... off the plant. (4)
As this damage (i.e. cutting flowers) is only temporary, the plant ...... (5)
i.e. The plant will continue to grow if we ...... off it.
i.e. The whole plant will not ______ if we cut flowers off it.
= If we cut flowers off the plant, the whole plant will NOT suffer.

(d) A plant has a root system, which is below the ground, and a shoot system, ... above ...... (6)
i.e. A plant has a ......, which is above the ground.
i.e. *The shoot system of a plant is above the ground.*

(e) The root takes in, or ______, water and minerals from the soil through the root hairs. (8)
i.e. The root absorbs ...... through the root hairs.
= Water and minerals ______ absorbed by the root through the root hairs.
Water and minerals are absorbed by the root through the root hairs.

(f) The roots hold, or ______, the plant firmly in position in the soil. (9)
i.e. The plant is anchored firmly in position in the soil by ......
the root = the root ______ (6)
∴ *The plant is anchored in the soil by the root system.*

(g) Both carrots and sugar beet are able to ...... (10)
By storing food in their roots they can ...... (11)
i.e. ...... can keep growing for more than one season.
∴ *Carrots can keep growing for more than one season.*

(h) Special bacteria live in the roots of plants such as ...... (12)
i.e. Clover and lucerne are examples of plants which have ...... living on their roots.
i.e. SOME plants have special bacteria which ......
= Special bacteria live on the roots of ......
Special bacteria live on the roots of SOME plants.
i.e. Special bacteria do NOT live on the roots of ALL plants.

(i) Clover and lucerne i.e. ______ plants are usually ploughed under the soil. (14)
By ploughing leguminous plants under the soil, the soil is made ...... (15)
i.e. The soil is made MORE fertile by ploughing leguminous plants such as ...... under the soil.
≠ The soil is made fertile by ploughing clover and lucerne under the soil.

EXERCISE A *Contextual reference*

1. In sentence 5, *such damage* refers to:
   (a) cutting flowers off the plant

(b) pruning the roots of the plant
(c) both cutting the flowers and pruning the roots

2. In sentence 8, *it* refers to:
(a) the shoot system
(b) the root of a plant

3. In sentence 11, *they* refers to:
(a) the roots of plants
(b) plants such as sugar beet and carrots

4. In sentence 13, *these simple forms of life* refers to:
(a) special bacteria
(b) legumes
(c) roots

EXERCISE B *Rephrasing*

Rewrite the following sentences replacing the words printed in italics with expressions from the text which have the same meaning.

EXAMPLE

The roots of plants *take in* water and minerals from the soil.
= The roots of plants *absorb* water and minerals from the soil.

1. *The single cells near the tip of each root* increase their surface area by extending outwards from the root.
2. The root *holds* the plant firmly in position in the soil.
3. Some plants have *simple forms of life* living on their roots.
4. We can improve the fertility of the soil by ploughing under *plants such as clover and lucerne.*
5. Sunlight provides the energy for the process of *converting water from the soil and carbon dioxide from the air into sugars and other carbohydrates.*
6. While growing, the seeds are protected by the *ripened ovary of the flower.*

EXERCISE C *Relationships between statements: consequence*

Study the following sentences:

Such damage is only temporary. The plant will continue to grow.

The relationship between the statements expressed in these two sentences is one of *consequence*. We can express this relationship in various ways:

(i) by joining the sentences together to make one sentence:

EXAMPLES

Such damage is only temporary *and so* the plant will continue to grow. (see text, sentence 5)

OR *As* / *Since* } such damage is only temporary, the plant will continue to grow.

(ii) by linking the two sentences as follows:

EXAMPLES

Such damage is only temporary. The plant will *consequently* / *therefore* / *thus* } continue to grow.

OR Such damage is only temporary. *Consequently,* / *Therefore,* / *Thus,* } the plant will continue to grow.

Relate the sentences indicated with the following expressions as in the examples above. Write out both sentences. Replace and re-order the words in the sentences where necessary.

(a) 10+11: (i) consequently (ii) thus (iii) and so (iv) since.

(b) 14+15: (i) thus (ii) and therefore (iii) consequently.

II LANGUAGE IN USE

EXERCISE A *Labelling a diagram*

Label the parts indicated in the diagram below of a mature bean plant using words and phrases from the text.

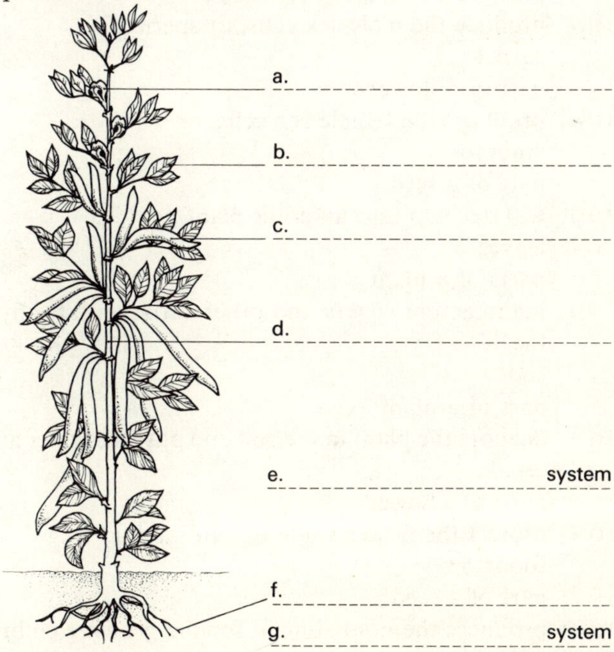

EXERCISE B *The definition of parts of a plant*

We can *define* different parts of a plant by

(a) naming them,
(b) stating the class they belong to,
(c) describing their function.

EXAMPLES

(i) NAME: root hairs
CLASS: parts of a plant
FUNCTION: absorb water and minerals from the soil
DEFINITION: *The* root hairs *are the* parts of a plant *which* absorb water and minerals from the soil.

(ii) NAME: stem
CLASS: part of a plant
FUNCTION: supports the shoot system
DEFINITION: *The* stem *is the* part of a plant *which* supports the shoot system.

Using the information below, write out complete definitions of each part of a plant as in the example above.

(a) NAME: fruit
CLASS: part of a plant
FUNCTION: protects the ripened, or matured, ovary

(b) NAME: stamens
CLASS: parts of a flower
FUNCTION: produce the male sex cells, or spermatia

(c) NAME: carpel
CLASS: part of a flower
FUNCTION: produces the female sex cells, or ovules

(d) NAME: embryo
CLASS: part of a seed
FUNCTION: will develop into an adult plant

(e) NAME: leaves
CLASS: parts of a plant
FUNCTION: manufacture sugars and other carbohydrates by photosynthesis

(f) NAME: root
CLASS: part of a plant
FUNCTION: anchors the plant in the soil and absorbs water and minerals

(g) NAME: sepals
CLASS: parts of a flower
FUNCTION: protect the flower while it is in the bud stage

(h) NAME: foodstore
CLASS: part of a seed
FUNCTION: produces the nourishment for the growing embryo

EXERCISE C *General statements of function*

We can make general statements about the function of different parts of a plant by naming them and saying what their function is but without saying what class they belong to.

EXAMPLE

NAME: root hairs
FUNCTION: absorb water and minerals from the soil
GENERAL STATEMENT: The root hairs absorb water and minerals from the soil.

Answer questions about what the following parts of a plant or flower or seed do by making general statements. Use the information from Exercise B, above.

	Statement of function
What do the root hairs do?	The root hairs absorb water and minerals from the soil.
OR What is the function of the root hairs?	OR *The function* of the root hairs *is to* absorb water and minerals from the soil.

(a) stem
(b) fruit
(c) stamens
(d) embryo
(e) leaves
(f) root
(g) sepals
(h) foodstore

III GRAMMAR

EXERCISE A *The forms of definitions*

Refer to Exercise B in Section II above.
Definitions often take one of the following forms:

1. [A] *is/are, may be defined as* [B] *which* [C].

   EXAMPLE

   [A The embryo] *is, may be defined as* [B the part of a flower] *which* [C will develop into another plant].
   The embryo *is* the part of a flower *which* will develop into another plant.
   OR The embryo *may be defined as* the part of a flower *which* will develop into another plant.

2. [B] *which* [C] *is/are called, is/are known as* [A].

   EXAMPLE

   [B The part of a flower] *which* [C will develop into another plant] *is called, is known as* [A the embryo].

The part of a flower *which* will develop into another plant *is called* the embryo.

OR The part of a flower *which* will develop into another plant *is known as* the embryo.

Expand the following into full definitions using the patterns illustrated above.

	A	B	C
1.	Photosynthesis	the process	transforms light energy from the sun into chemical energy.
2.	A soil profile	a succession of soil horizons	extends from the surface of the soil to the parent rock.
3.	Aerobic bacteria	organisms	can live in the presence of air.
4.	Osmosis	biophysical process	takes place through the tissues of living plants.
5.	A leaf	complex structure	utilizes energy from the sun in the manufacture of food.
6.	Chloroplasts	bodies	absorb sunlight and manufacture food.
7.	Stomata of plants	minute openings on the surface of a leaf	lead to the interior of the leaf and the chloroplasts.
8.	Chlorophyll	the chemical	enables sunlight to convert carbon dioxide into food and other substances.

EXERCISE B *The impersonal passive*

Examine the following active and passive sentences, and note that passive sentences contain some form of the verb *to be* together with a past participle:

Active	*Passive*
We *improve* the fertility of the soil.	The fertility of the soil *is improved.*
The fruit *encloses* the seeds.	The seeds *are enclosed* by the fruit.
We *should plough under* leguminous plants.	Leguminous plants *should be ploughed under.*

Notice that the passive brings into prominence the thing to which the action is being done by placing it at the beginning of the sentence.

Active	*Passive*
The botanist defines a fruit as a ripened ovary.	A fruit is defined (by the botanist) as a ripened ovary.
The stem supports the plant.	The plant is supported by the stem.

Note that the words in brackets are optional, and are often omitted in scientific writing.

Write down the passive version of the following active sentences. Then combine the passive sentences you have written following the clues provided.

EXAMPLE

Active: The tiny root hairs absorb water and minerals.
Passive: Water and minerals *are absorbed* by the tiny root hairs.
Active: An increase in the number of root hairs increases the power of absorption.
Passive: The power of absorption *is increased* by an increase in the number of root hairs.

Water and minerals *are absorbed* by the tiny root hairs. Therefore, the power of absorption *is increased* by an increase in the number of root hairs.

1. Active: The fruit encloses the seeds.
   Passive: The seeds ...... by the fruit.
   Active: The fruit protects them while they are developing.
   Passive: They ...... by the fruit while they are developing.

The seeds ...... by the fruit. Consequently, they ...... while they are developing.

2. Active: The plant takes in oxygen.
   Passive: Oxygen is ...... in by the plant.
   Active: The plant uses oxygen to break down carbohydrates.
   Passive: Oxygen is ...... by the plant to break down carbohydrates.

Oxygen ...... in by the plant and ...... to break down carbohydrates.

3. Active: Wind and insects transfer pollen from one flower to another.
   Passive: Pollen ...... by wind and insects from one flower to another.
   Active: They deposit the pollen on the stigma of the other flower.
   Passive: The pollen ...... on the stigmas of the other flower.

When pollen ...... by the wind and insects from one flower to another, it ...... on the stigmas of the other flower.

4. Active: The human body requires small quantities of several minerals.
   Passive: Small quantities of several minerals ...... by the human body.
   Active: The human body obtains these minerals from plants.
   Passive: These minerals ...... by the human body from plants.

Small quantities of several minerals which are ...... by the human body ...... from plants.

5. Active: We can use some roots to reproduce the species.
   Passive: Some roots ...... to reproduce the species.
   Active: We should remove the whole root of harmful weeds such as docks instead of ploughing them in lightly.
   Passive: The whole root of harmful weeds such as docks ...... instead of ...... in lightly.

Some roots . . . . . . to reproduce the species. Thus, the whole root of harmful weeds such as docks . . . . . . instead of . . . . . . lightly.

6. Active: Soil texture influences all aspects of root development.
   Passive: All aspects of root development . . . . . . by soil texture.
   Active: A heavy compact soil creates a physical barrier to root growth.
   Passive: A physical barrier to root growth . . . . . . by a heavy compact soil.

All aspects of root development . . . . . . by soil texture. For example, a physical barrier to root growth . . . . . . by a heavy compact soil.

7. Active: Too much cultivation destroys the soil structure.
   Passive: Soil structure . . . . . . by too much cultivation.
   Active: A pasture phase under grass can improve the structure.
   Passive: The structure . . . . . . by a pasture phase under grass.

Since soil structure . . . . . . by too much cultivation, the structure . . . . . . by a pasture phase under grass.

8. Active: The plant manufactures food from chemical substances present in the soil and air.
   Passive: Food . . . . . . by the plant from chemical substances present in the soil and air.
   Active: The roots take in chemical substances from the soil.
   Passive: Chemical substances . . . . . . in from the soil by the roots.
   Active: The leaves take in carbon dioxide from the air.
   Passive: Carbon dioxide is . . . . . . from the air by the leaves.

Food is . . . . . . by the plant from chemical substances . . . . . . from the soil by the roots and from carbon dioxide . . . . . . from the air by the leaves.

9. Active: We can use a unit called a soil profile to describe soils.
   Passive: A unit called a soil profile . . . . . . to describe soils.
   Active: When we wish to compare two soils, we examine their profiles.
   Passive: When we wish to compare two soils, their profiles . . . . . .
   Active: We can define a soil as having an individual profile.
   Passive: The soil can be . . . . . . as having an individual profile.

A unit called a soil profile . . . . . . to describe soils. So when we wish to compare two soils, their profiles . . . . . ., and each soil can thus . . . . . . as having an individual profile.

10. Active: Ploughing 'turns in' the whole surface of a field.
    Passive: The whole surface of the field is '. . . . . .' by ploughing.
    Active: It buries and kills the weeds.
    Passive: The weeds . . . . . . and . . . . . .
    Active: It loosens and exposes the soil to the air.
    Passive: The soil . . . . . . and . . . . . . to the air.

The whole surface of the field '. . . . . .' by ploughing, with the result that weeds . . . . . . and . . . . . . and the soil . . . . . . and . . . . . . to the air.

IV SUMMARY CHECK

Complete the following text by filling in the blank spaces with the expressions given below. A dotted line . . . . . . requires a phrase to be added and a straight line _____ requires a word to be added.

roots	shoot system	soil
soil air	carbon dioxide	photosynthesis
made up of	ripened ovary	water and minerals
seed	living	such as
function	specialized	more fertile
by	reproductive organs	consists of
their roots	organic materials	process
conversion	are produced	carbohydrates

A plant is a living organism . . . . . . different parts each of which has a _____ function. The basic parts of a plant are the root system and the . . . . . .

The root absorbs water and minerals from the _____. Plants such as sugar beet and carrots store food in . . . . . . Leguminous plants . . . . . . clover and lucerne have special bacteria _____ on their roots which take nitrogen out of the . . . . . . Consequently, when they are ploughed under, the soil is made . . . . . .

The shoot system . . . . . . the stem, the leaves, flowers and fruit. An important _____ of the stem is to enable . . . . . . to pass up to the leaves and flowers and . . . . . . such as sugar to travel down to the _____. In the leaves _____ takes place. The process results in the _____ of water from the soil and . . . . . . from the air into sugars and other _____. During the _____ oxygen is formed and released into the air. The plant's . . . . . . are contained in the flower. The spermatia . . . . . . by the stamens and the ovules are produced _____ the carpel. The fruit, the . . . . . . of the flower, encloses and protects the _____.

2 The Life Cycle of a Plant

I READING AND COMPREHENSION

[1]The life cycle of a typical annual plant can be divided into several stages. [2]The first stage is germination. [3]Seeds remain dormant, or in a resting state, if they are kept cool and dry. [4]When the amount of moisture and the temperature level are right, the seeds germinate and start growing.

[5]Certain conditions are necessary for this to happen. [6]An essential condition is that the seeds must be alive. [7]Sometimes seeds are dried at a temperature which is too high. [8]This has two effects: the water content in the seeds is reduced too much, and certain essential proteins are destroyed. [9]As a result, the seeds die.

(a) Before a seed germinates it is in a dormant state.
(b) When the temperature level is right a seed will germinate.
(c) If seeds are dried at too high a temperature they will die.

[10]Other conditions for germination concern the amount of moisture in the soil. [11]If dry seeds are planted in a dry soil, they will not germinate until it rains. [12]On the other hand, if there is too much water in the soil, the seeds will not germinate either. [13]This is because wet soils remain cold for a longer period of time than drier, well-drained soils. [14]If the soil is too cold germination will not occur. [15]An additional reason for seeds not germinating is that badly drained soils may lack sufficient oxygen. [16]Dormant seeds require very little oxygen in order to stay alive, but when they start to germinate they require more.

(d) If a soil is too dry seeds will not germinate.
(e) The temperature of wet soils is lower than that of well-drained soils.
(f) Dormant seeds cannot stay alive in a badly drained soil.

[17]In the first stage of germination the primary root, or radicle, emerges. [18]Then the stem pushes its way upward until it appears above the surface of the soil. [19]At the same time the root system grows downward, and begins to spread through the soil. [20]In the early stages of development the seedling

depends entirely on the foodstore in the seed but as soon as the first leaves are produced, it is able to manufacture food for itself. [21]The seedling begins photosynthesis.

(g) The root system forms after the stem appears above the surface of the soil.

(h) The seed contains enough food to nourish the seedling until the first leaves are produced.

[22]Next, the plant enters the stage of rapid growth. [23]In this stage of the life cycle, the plant begins to grow to its full size. [24]When it is mature enough, it flowers, and when this happens pollination and fertilization are ready to take place. [25]In the process of pollination the pollen is carried by wind or insects from the stamens to the stigma of the carpel. [26]It germinates on the stigma and grows down the style into the ovary, where fertilization takes place.

Solutions

(a) If seeds are kept cool and dry they remain _____. (3)
Seeds germinate when . . . . . . (4)
i.e. Seeds are in a dormant state UNTIL the amount of moisture and the temperature level are right.
i.e. Seeds are . . . . . . BEFORE they _____.
= *Before a seed germinates it is in a dormant state.*

(b) Seeds germinate and start growing when
(i) the amount of _____ is right,
(ii) the _____ level is right. (4)
i.e. When the temperature level is right BUT the amount of moisture is NOT right, the seed will NOT _____.
i.e. If ONLY the temperature level is right the seed will NOT germinate.

(c) Sometimes the temperature is . . . . . . when seeds are dried. (7)
As a result, the seeds _____. (9)
i.e. When seeds . . . . . . at too high a _____ they will die.
= *If seeds are dried at too high a temperature they will die.*

(d) Dry seeds will not germinate if they are planted in . . . . . . (11)
If a soil is dry there is NOT ENOUGH _____ in it.
If there is NOT ENOUGH water in the soil, the seeds will not _____.
= If the soil is TOO _____ the seeds will not _____.
If a soil is too dry the seeds will not germinate.

(e) We are comparing _____ soils with _____, _____-_____ soils. (13)
Wet soils . . . . . . LONGER than drier, well-drained soils. (13)
If wet soils remain cold LONGER than well-drained soils, the _____ of wet soils is lower than that of . . . . . .
i.e. *The temperature of wet soils is lower than that of well-drained soils.*

(f) A soil which is badly drained may not have ...... for seeds to _____. (15)
i.e. Seeds cannot germinate in a soil which lacks sufficient oxygen.
but In order to ...... dormant seeds require very little _____. (16)
∴ Dormant seeds CAN stay alive in a badly drained soil, but they CANNOT germinate.

(g) The *first stage* of germination is the emergence of the ...... (17)
Then, the stem pushes upward until ...... (18)
i.e. BEFORE the stem appears above the surface of the soil, the primary root has emerged i.e. has formed.
∴ It is NOT TRUE to say that the root system forms AFTER the stem appears above the surface of the soil.

(h) The seedling depends on ...... in the seed in the early stage of development. (20)
The early stage of development *i.e.* before the ...... (20)
∴ Until the first leaves are produced, the seedling depends on ......
i.e. The foodstore in the seed contains enough _____ to nourish the _____.
The foodstore in the seed contains enough food to nourish the seedling.

EXERCISE A *Contextual reference*

1. In sentence 5 *this* refers to:
   (a) the life cycle of a plant
   (b) the germination of a seed
   (c) the right temperature level
2. In sentence 8 *this* refers to:
   (a) too high a temperature
   (b) the drying of seeds
   (c) the condition that seeds must be alive
3. In sentence 13 *this* refers to:
   (a) too much water in the soil
   (b) seeds not germinating
4. In sentence 20 *it* refers to:
   (a) the seed
   (b) the foodstore
   (c) the seedling
5. In sentence 24 *it* ...... *it* refers to:
   (a) its full size
   (b) the plant
   (c) the life cycle

6. In sentence 24 *this happens* refers to:
   (a) when the plant begins to grow to its full size
   (b) when the plant enters the stage of rapid growth
   (c) when the plant flowers
7. In sentence 26 *it* refers to:
   (a) the pollen
   (b) the stamens
   (c) the stigma

EXERCISE B *Rephrasing*

Rewrite the following sentences replacing the words printed in italics with expressions from the text which have the same meaning. (Refer to Unit 1, Reading and Comprehension, Exercise B.)

1. The seed *starts growing* when there is enough air or water and the temperature is right.
2. A seed will only germinate when there is *enough* air in the soil.
3. Seeds which are *in a resting state* require very little air to remain alive.
4. As soon as the stem and leaves appear above the surface of the soil, they begin *to manufacture food.*
5. After the plant has appeared above the surface of the soil it enters *the stage of life when it begins to grow to its full size.*
6. *The process of carrying the pollen to the stigma* is brought about by wind or insects.

EXERCISE C *Relationships between statements*

Part 1 *Contrast*

Study the following sentences:

Seeds *remain dormant* if they are kept cool and dry.
When the amount of moisture and the temperature level are right, the seeds *germinate and start growing.*

The relationship between these statements is one of *contrast.* The italicized words indicate the contrast. We can relate these sentences as follows:

EXAMPLES

Seeds remain dormant if they are kept cool and dry. *However,* when the amount of moisture and the temperature level are right, the seeds germinate and start growing.

OR Seeds remain dormant if they are kept cool and dry. When the amount of moisture and the temperature level are right, *however*, the seeds germinate and start growing.

Rewrite the following sentences by using *however*. Replace and re-order the words in the sentences where necessary.

(a) 11+12
(b) 16
(c) 20
(d) Unit 1; 3+4

Part 2 *Consequence*
Relate the sentences indicated with the following expressions as in Unit 1 Reading and Comprehension Exercise C. Write out both sentences and replace and re-order the words in the sentences where necessary.

(a) 7+8: (i) as a consequence (ii) and so (iii) consequently
(b) 8+9: (i) hence (ii) consequently (iii) therefore.

II LANGUAGE IN USE

EXERCISE A *Definitions of process*

Processes can be defined by identifying them by what happens during their course of action.

EXAMPLE

NAME OF PROCESS: photosynthesis

IDENTIFYING DESCRIPTION: water and carbon dioxide are built up to form sugars and other carbohydrates in the presence of light

DEFINITION: Photosynthesis *is* the process *whereby* water and carbon dioxide are built up to form sugars and other carbohydrates in the presence of light.

OR: Photosynthesis is the process *by which* water and carbon dioxide are built up to form sugars and other carbohydrates in the presence of light.

Match the names of the processes in the list on the left to the identifying descriptions of their course of action in the list on the right. Write definitions of the process in the two ways shown above in the example.

NAME OF PROCESS		DESCRIPTION OF COURSE OF ACTION
(a) transpiration	(i)	one of the male gametes unites with the female gamete in the ovule
(b) germination	(ii)	pollen grains are transferred from the stamen to the stigma of the female parts.
(c) pollination	(iii)	water passes through the leaf cells and evaporates into the air
(d) fertilization	(iv)	nutrients in the soil pass through the cell membranes into the root hairs
(e) osmosis	(v)	the seed 'awakens' from its dormant state and starts growing

EXERCISE B *General statements of process*

We can make general statements about processes by describing what happens during their course of action.

EXAMPLE

NAME OF PROCESS: photosynthesis

GENERAL DESCRIPTION: water and carbon dioxide are built up to form sugars and other carbohydrates in the presence of light

GENERAL STATEMENT: *In the process of* photosynthesis water and carbon dioxide are built up to form sugars and other carbohydrates in the presence of light.

Answer questions about the following processes using the information from Exercise A above.

EXAMPLE

	STATEMENT OF PROCESS
What happens in/during the process of photosynthesis?	*In/during the process of* photosynthesis water and carbon dioxide are built up to form sugars and other carbohydrates in the presence of light.

(a) transpiration
(b) germination
(c) pollination
(d) fertilization
(e) osmosis

EXERCISE C *Statements of function and process*

Statements of function (see Unit 1 Language in Use, Exercise C) may be followed by statements of process.

EXAMPLE

STATEMENT OF FUNCTION:

NAME OF PLANT PART: root hairs
FUNCTION: absorb water and minerals from the soil

STATEMENT OF PROCESS:

NAME OF PROCESS: osmosis
COURSE OF ACTION: nutrients in the soil pass through the cell membranes into the root hairs

= *The function of* the root hairs *is to* absorb water and minerals from the soil. *The process whereby this is done is called* osmosis. *In this process* nutrients in the soil pass through the cell membranes into the root hairs.

Write out short paragraphs as in the example above using the information below, and the words in brackets.

(a) NAME OF PLANT PART: leaves
FUNCTION: manufacture food for the plant
NAME OF PROCESS: photosynthesis
COURSE OF ACTION: water from the soil and carbon dioxide from the air are built up to form sugars and other carbohydrates in the presence of light.

(b) NAME OF PLANT PART: embryo in the seed
FUNCTION: develop into an adult plant
NAME OF PROCESS: (the first stage in) germination
COURSE OF ACTION: the seed 'awakens' from its dormant state and starts growing

(c) NAME OF PLANT PART: flower
FUNCTION: bring about the reproduction of the plant
NAME OF PROCESS (1): (the first stage in) pollination
COURSE OF ACTION: pollen grains are transferred from the stamens to the stigma of the female parts
NAME OF PROCESS (2): (the next stage is called) fertilization
COURSE OF ACTION: one of the male gametes from the pollen unites with the female gamete in the ovule

EXERCISE D *Definitions and descriptions of processes*

Definitions of processes are often followed by statements describing the various stages of the processes. Frequently the description is accompanied by a labelled diagram.

Part 1 *Labelling of diagrams*

Complete the labelling of the following diagrams by inserting the correct labels from the list of words and phrases below.

1.

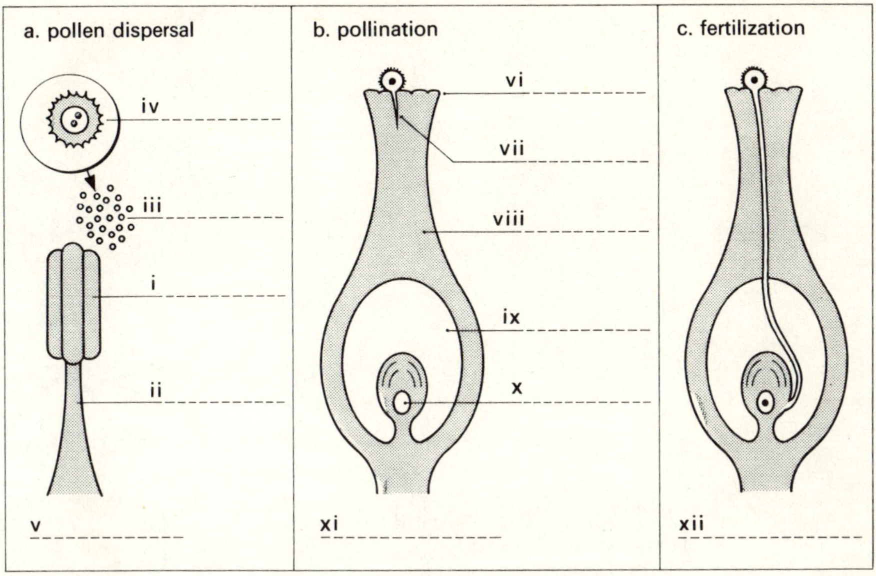

FIG. 1 *Reproduction in a plant*

pollen grain	stigma	carpel
anther	stamen	pollen grain germinating
ovary	ovule	union of male and female gametes
filament	pollen	style

2.

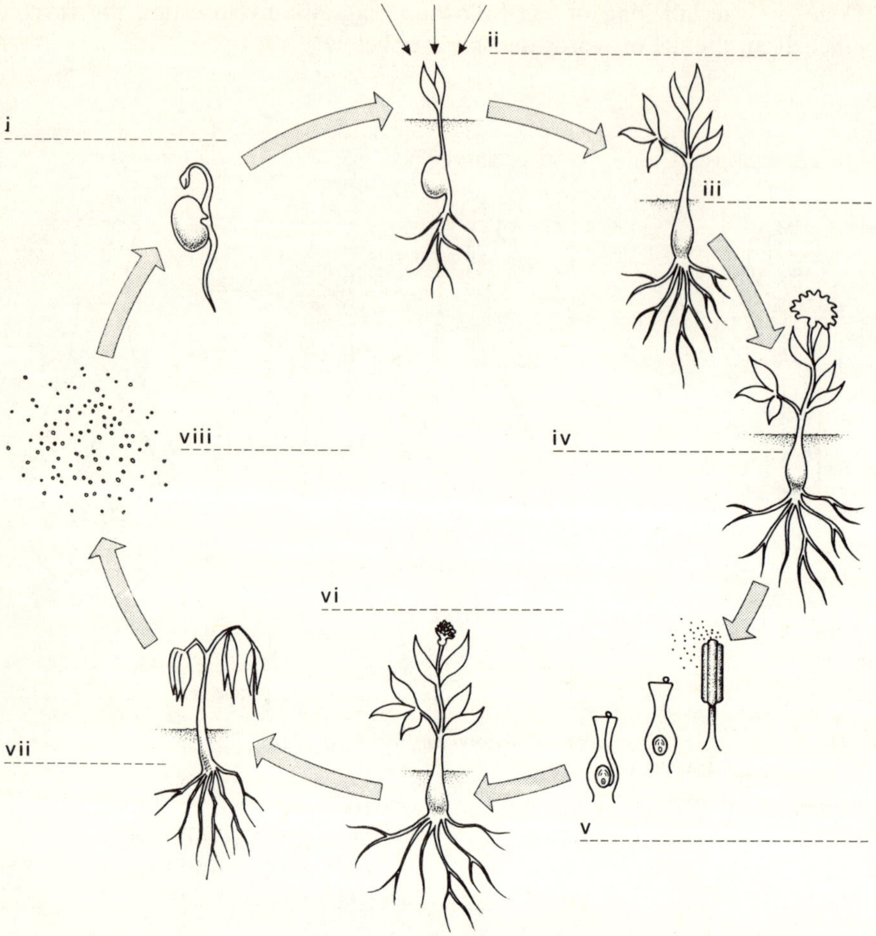

FIG. 2 *The life cycle of a plant*

fruit and seed production	decay of vegetative parts
photosynthesis begins	pollination and fertilization
plant flowers	stage of rapid vegetable growth
seed dispersal	germination

3.

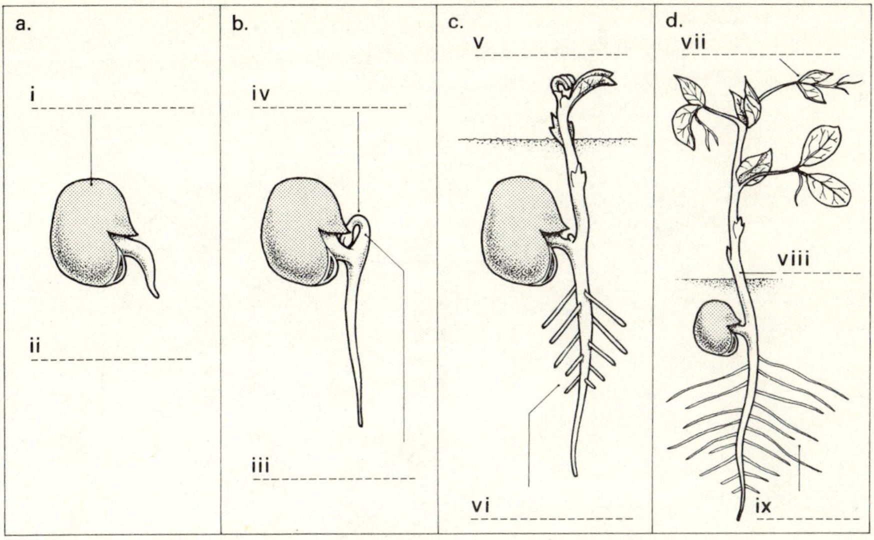

FIG. 3 *The germination of a broad bean*

secondary roots develop
root system spreads through soil
split testa
photosynthesis can begin
plumule
radicle
curved to protect growing point
leaves sprouting
main shoot

4.

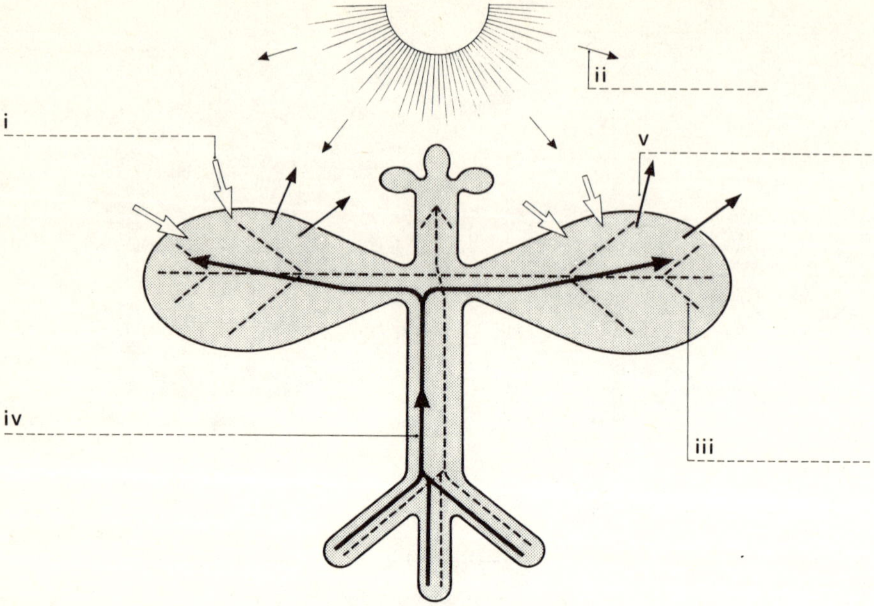

FIG. 4 *Photosynthesis*

water from soil
sunlight
sunlight used to assist combination of CO_2 and H_2O
carbon dioxide taken in
oxygen given off as gas

Part 2

Arrange the sentences below in the order which corresponds to the stages of the processes illustrated in the diagrams above. Place the definition at the beginning of each description.

EXAMPLE

1. It is carried by wind or insect from the stamens to the stigma of the carpel.
2. Pollination is the process whereby pollen grains are transferred from the male parts of the flower to the female parts.
3. The next stage is fertilization.
4. The male sexual organ, or stamens, produces the pollen which contains two male gametes.

(2) Pollination is the process whereby pollen grains are transferred from the male parts of the flower to the female parts. (4) The male sexual organ, or stamens, produces the pollen which contains two male gametes. (1) It is carried by wind or insect from the stamens to the stigma of the carpel. (3) The next stage is fertilization.

1. It fertilizes the female gamete in the ovule by uniting with it.
   Fertilization is the process whereby one of the male gametes from the pollen unites with the female gamete in the ovule.
   One of the male gametes goes down the style into the ovary chamber.
   The pollen carrying the male gametes germinates on the stigma.
2. The first stage of plant growth is the germination of the seed. This forms the beginnings of the stem and root systems.
   Fruit and seeds are produced.
   As soon as the stem appears above the ground food manufacture, or photosynthesis, begins.
   This is followed by the decay of the vegetative parts.
   The life cycle of a plant is the period of time during which the plant grows from a seed, flowers and dies.
   When the plant flowers it is ready for pollination and fertilization.
   The seeds are dispersed by insects, animals or wind, and the plant dies.
   Pollen is transferred from the stamens to the stigma where it germinates.
   The plant enters the period of rapid growth. During this time the vegetative parts grow to full size.
3. When the young plant breaks the soil surface, food manufacture by photosynthesis can begin.
   Germination is the process whereby the seed 'awakens' from its dormant state and begins to grow.
   The main shoot grows upwards sprouting leaves; the root system spreads through the soil.
   The first stage in the germination of a bean is the splitting of the testa. The radicle emerges from the testa and starts to grow downwards.
   Below the soil surface secondary roots develop.
   The plumule, which is curved to protect the growing point, begins to grow up towards the light.
4. The sunlight provides the energy to bind CO_2 and H_2O together to form sugars and other carbohydrates.
   Photosynthesis is the process whereby the plant manufactures food for itself.
   Carbohydrates go to the growing points in the plant, enlarging tissues.
   Oxygen is given off as a gas.
   We may summarize this chemical process as follows:
   $6CO_2 + 6H_2O \rightarrow C_6H_{12}O_6 + 6O_2$
   This is combined with water from the soil in the presence of sunlight.
   Carbon dioxide is taken in through the leaf cells.

Part 3

In descriptions of processes, words or phrases such as *first(ly), then, the next step/stage, subsequently, finally* etc. are frequently used to introduce each stage.

EXAMPLE

1. Pollination is the process whereby pollen grains are transferred from the male parts of the flower to the female parts. *Firstly*, the male sexual organ, or stamens, produces the pollen which contains two male gametes. *Then* the pollen is carried by wind or insect from the stamens to the stigma of the carpel. *The next stage* is fertilization.

Rewrite the paragraphs from Part 1 above placing the following expressions in the appropriate sentences.

1. firstly, then, finally
2. after that, during this stage, next, finally
3. next, also at this stage, finally, at the same time
4. first of all, subsequently, at the same time

III GRAMMAR

EXERCISE A *Time expressions*

Part 1 *after, before, when, as soon as, while*

Look at these sentences:

(i) *First* the seed is provided with water, warmth and air, then it starts to germinate. (after, until, while)
(ii) *After* the seed is provided with water, warmth and air, it starts to germinate.

Now rewrite the following sentences in the same way, choosing one of the time expressions given in brackets and putting it *at the beginning of the sentence* as indicated (/). Omit the words in italics.

1. /The seedling begins to manufacture food for itself. *But first* it uses up the food stored in the seed. (when, before, after)
2. /The young shoot appears above the surface of the ground. *Then* it begins the process of photosynthesis. (before, as soon as, while)
3. *Once*/ the oxygen has combined with and broken down the various complex sugars, energy is released. (before, after, while)
4. /Dormant seeds are inactive. *During this time* they use very little air. (when, before, while)
5. /The young rice plants are transplanted to the paddy fields. *But first* they are grown in nurseries for a few weeks where proper care can be given to the seedlings. (before, while, after)

6. *Once*/ the shoot appears, the plant *then* grows both above and below the ground. (before, while, after)
7. *During the time*/ the seedlings are small, there are few leaves present to use sunlight for photosynthesis. (while, before, as soon as)
8. /A crop of nitrogen-fixing legumes was ploughed in. *As a result* the next crop produced a higher yield. (while, until, after)
9. /The spores of disease organisms land on the plant. *At the same time* they are killed by the fungicide which has been sprayed or dusted on to the plant surfaces. (as soon as, while, before)
10. /*Sometimes* there is too much water in the soil. *On these* occasions it must be drained off. (after, until, when)

Part 2 *until*

Look at this sentence:

(i) After the seed is provided with water, warmth and air it starts to germinate.

We can state this in a negative form, as follows:

(ii) The seed does NOT start to germinate UNTIL it is provided with water, warmth and air.

Now look at the sentences you have written in Part 1 and rewrite the following as in (ii) above, using *until* and making the other changes indicated:

Sentence 1. (change *uses* to *has used*)
Sentence 2. (change *appears* to *has appeared*)
Sentence 3. (no other change)
Sentence 4. (change *inactive* to *active*)
Sentence 8. (change the *next crop* to *the soil*)

Part 3 *then, during, throughout, prior to, first*

Compare the following sentences with your answers to Exercise A, Part 1. If the sentences have approximately the same meaning put a tick in the box, if not, put a cross. The first two have been done for you.

1. Prior to the seedling manufacturing food for itself, it uses up the food stored in the seed. ☑
2. The process of photosynthesis begins and then the young shoot appears above the surface of the ground. ☒
3. Oxygen combines with and breaks down the various complex sugars prior to energy being released. ☐
4. Throughout the time that dormant seeds are inactive they use very little air. ☐
5. The young rice plants are transplanted to the paddy fields and then they are grown in nurseries for a few weeks where proper care can be given to the seedlings. ☐

6. Prior to the shoot appearing the plant grows above and below the ground. ☐
7. There are few leaves present to use sunlight for photosynthesis through the time that the seedlings are small. ☐
8. During the time nitrogen fixing legumes were ploughed in, the next crop produced a higher yield. ☐
9. If the spores of disease organisms land on the plant they are killed by the fungicide which has been sprayed or dusted on to the plant surfaces. ☐
10. If there is too much water in the soil it must be drained off. ☐

EXERCISE B *Expressions of degree*

Part 1 *too*

Look at the following pairs of sentences. Notice that the first half of each sentence describes *a state* and the second half expresses *a consequence*.

(i) The soil was dry so the seed could not germinate.
(ii) The soil was heavy and clayey and, as a result, it was unsuitable for root crops.

We can express these sentences in another way using the expression of degree *too*:

(iii) The soil was *too dry* for the seed to germinate.
(iv) The soil was *too heavy and clayey* to be suitable for root crops.

Rewrite the following sentences using *too* as in examples (iii) and (iv) above, and make any other changes that are necessary.

1. The soil particles are fine so the water cannot percolate easily through the soil.
2. Because the soil was compact, it was not suitable for root crops.
3. Soil aeration was inadequate and consequently the plant could not receive a proper supply of oxygen.
4. As a result of the land being waterlogged, it was not possible to produce a healthy crop.
5. The soil profile was so shallow that it could not give the roots sufficient anchorage.
6. As the root system was poorly developed, the plant could not produce a good top growth.
7. It was dark and as a result photosynthesis could not take place.
8. The current was swift so that the silt would not be deposited.

Part 2 *enough*

Now look at the sentences you have written and rewrite them using *not* + *adjective* + *enough*, using the adjective given for each sentence.

EXAMPLES

(i) The soil was too dry for the seed to germinate. (wet)

= The soil was *not wet enough* for the seed to germinate.

(ii) The soil was too heavy and clayey to be suitable for root crops. (light and sandy)

= The soil was *not light and sandy enough* to be suitable for root crops.

1. coarse; 2. loose; 3. adequate; 4. well-drained; 5. deep; 6. well-developed; 7. bright; 8. slow.

IV SUMMARY CHECK

Complete the following text by filling in the blank spaces with the expressions given below. A dotted line ...... requires a phrase to be added and a straight line _____ requires a word.

root system	rapid growth	may be reduced
colder	too high	sufficient air
seeds	too much	temperature level
temperature	testa	food manufacture
germination (3 times)	photosynthesis	well-drained soils
downwards	little moisture	dependent on the foodstore
not	in the soil	secondary roots
life	water and air	surface of the soil
up	destroyed	mature

The first stage in the life cycle of a plant is _____. Certain conditions are necessary for _____ to occur. Firstly, the _____ must be alive. If seeds are dried at ...... a temperature, the water content in the seeds ...... too much and certain essential proteins _____. Secondly, the amount of moisture in the soil must be right. If there is too ...... in the soil, seeds will not germinate. However, if there is ...... water in the soil, seeds will _____ germinate either because wet soils tend to be _____ than drier, ... This is the third condition necessary for germination to occur. The _____ of the soil must be right. A fourth condition concerns the amount of air ...... A wet, badly drained soil may lack ...... for seeds to germinate. Thus, we may say that _____ only happens under the right conditions; when there is _____, sufficient ......, and the right _____.

The first stage in the germination of, for example, a bean is the splitting of the _____. The radicle emerges and starts to grow _____. Next, the curved plumule begins to grow _____ towards the light. Meanwhile, the ...... is beginning to spread through the soil. In these early stages of development, the seedling is entirely ...... in the seed. After the young plant has broken the ...... and the first leaves are produced ...... by _____ can begin. By this time, below the soil surface ...... are developing. The plant is ready to begin the stage of ...... during which it grows to its full _____ size.

3 The Origin and Composition of Soil

I READING AND COMPREHENSION

[1]Soil is a residue composed of two main ingredients: mineral material and organic material. [2]Organic material originates from dead plants and animals and materials other than this are derived from rocks of various kinds. [3]These rocks are broken down into small particles by mechanical disintegration and chemical decomposition. [4]This breaking down process, known as weathering, may thus be both physical and chemical.

[5]When weathering processes are largely physical – by heat or wind, for instance – the composition of the soil is very similar to that of the parent rock. [6]In arid regions weathering is mostly by physical means. [7]But in humid regions chemical processes of weathering are equally important. [8]In such regions rock particles are affected by water which may contain carbonic or other weak acids. [9]These acids dissolve some of the particles in the rocks. [10]The mineral material that is left behind is insoluble. [11]Consequently, the insoluble mineral residues in the soils have less resemblance to the original rocks. [12]There are larger amounts of organic matter in the soil, too.

(a) Mineral materials in soil are derived from various kinds of rocks.
(b) Physical and chemical weathering breaks down rocks into small particles.
(c) The composition of the soil in arid regions has a close resemblance to that of the parent rock.
(d) Soils of humid regions are similar to the parent rock.

[13]The process of soil formation results in the development of the soil profile. [14]This is made up of a succession of horizontal layers, or 'horizons', of varying thickness, from the surface to the parent rock. [15]Generally speaking, there are three distinct horizons, known as A, B and C. [16]A is the top soil, which is coarse-grained, and dark in colour because of the presence of humus. [17]B is known as the sub-soil which contains some of the products leached, or washed, out of the A horizon. [18]The C horizon consists of parent material which has been weathered in the upper part, and unweathered rock below.

(e) The succession of soil horizons from the surface to the parent rock is known as the soil profile.

[19]Any sample of soil contains particles of different sizes. [20]These have been divided into the following size groups:

TABLE 1

Material	*Diameter (mm)*
gravel	more than 2·0
coarse sand	2·0–0·2
fine sand	0·2–0·02
silt	0·02–0·002
clay	less than 0·002

[21]Soils range from pure clays to pure sands. [22]Most of them contain various proportions of sand, silt and clay and these varying proportions make up a soil's textural class. [23]The principle classes in order of increasing fineness of material are sand, loamy sand, loam, silt loam, silty clay loam, clay loam, silt and clay.

(f) Particles smaller than silt are clay particles.
(g) A clay loam contains particles of greater fineness than a loamy sand.

[24]Any soil contains both mineral and organic matter. [25]Clay particles are the most important of the mineral particles because they are the smallest. [26]Smaller sized particles have a greater exposed surface area than larger sized particles. [27]The smaller the size of a particle, the greater is its reactivity. [28]That is to say, smaller sized particles can react or combine with water, nutrients and humus more easily than larger sized particles. [29]Thus, a clay soil is more reactive than any other type of soil. [30]Humus from decomposed organic matter is vital to a soil as it makes a heavy soil lighter. [31]In addition, it helps to bind the mineral particles together in 'crumbs'.

Solutions

(a) Soil is composed of mineral material and . . . . . . (1)
i.e. If the soil material is not organic, it is derived from . . . . . . (2)
Rocks of various kinds i.e. _____ materials.
∴ Mineral materials in soil . . . . . . various kinds of rocks.
Mineral materials in soil are derived from various kinds of rocks.

(b) Rocks of various kinds are . . . . . . into small particles. (3)
This breaking down process is known as _____. (4)
Weathering may be BOTH _____ AND _____. (4)
∴ Rocks are . . . . . . into . . . . . . by physical and chemical weathering.
= *Physical and chemical weathering breaks down rocks into small particles.*

(c) In arid regions the weathering processes are mostly _____. (6)
If the weathering processes are physical the composition of the soil is very _____ to the composition of the _____ rock. (5)
∴ The composition of the soil in . . . . . . is very similar to that of the parent rock.
is very similar to = has a close _____ to (11)
∴ The . . . . . . in arid regions has a close resemblance to that of . . . . . .
The composition of the soil in arid regions has a close resemblance to that of the parent rock.

(d) We are comparing the soils of arid regions with those of . . . . . . (7)
In humid regions the water dissolves some of the rock particles, but leaves behind mineral material that is _____. (10)
These insoluble mineral _____ have . . . . . . to the original (i.e. the parent) rock. (11)
have less resemblance to = ARE LESS SIMILAR TO (see solution **c**)
∴ the residues in the soils of humid regions are LESS SIMILAR TO the . . . . . . than the soils of arid regions.
∴ It is NOT TRUE to say that the soils of humid regions are similar to the parent rock.

(e) From the surface to the parent rock there are a succession of horizontal layers, or _____. (14)
The succession of soil horizons makes up the soil _____. (13)
∴ The soil profile is the succession of soil horizons from . . . . . .
= *The succession of soil horizons from the surface to the parent rock is known as the soil profile* (i.e. a form of definition. See Unit 1 Grammar).

(f) Silt: 0·02 – 0·002 mm.
Clay: . . . . . . 0·002 mm. (Table 1)
If clay particles are less than 0·002 mm. diameter they are SMALLER than . . . . . .
∴ *Particles smaller than silt are clay particles.*

(g) Sand, loamy sand, loam, silt loam, silty clay loam, clay loam, silt and clay are arranged in order of . . . . . . of particles. (23)
increasing fineness of particles = *greater fineness* of particles
Clay loam comes after _____ sand.
∴ A clay loam contains . . . . . . than a loamy sand.
A clay loam contains particles of greater fineness than a loamy sand.

EXERCISE A *Contextual reference*

1. In sentence 2 *this* refers to:
   (a) mineral material
   (b) organic material

2. In sentence 5 *that* refers to:
   (a) heat or wind
   (b) the composition
   (c) the soil
3. In sentence 8 *such regions* refers to:
   (a) arid regions
   (b) humid regions
4. In sentence 14 *this* refers to:
   (a) the process of soil formation
   (b) the development of the soil profile
   (c) the soil profile itself
5. In sentence 20 *these* refers to:
   (a) particles
   (b) different sizes
6. In sentence 22 *them* refers to:
   (a) soils
   (b) pure clays
   (c) pure sands
7. In sentence 25 *they* refers to:
   (a) mineral particles
   (b) clay particles
8. In sentence 31 *it* refers to:
   (a) a heavy soil
   (b) organic matter
   (c) humus

EXERCISE B *Rephrasing*

Rewrite the following sentences replacing the words printed in italics with expressions from the text which have the same meaning.

1. *Material other than mineral material* is derived from dead plants and animals.
2. *Breaking down rocks into small particles* is performed mostly by *heat or wind* in arid and semi-arid regions.
3. *The remains of mineral materials that cannot be dissolved in water* have little *similarity* to the parent rocks in humid regions.
4. *The succession of horizontal layers* in a soil are called the top soil, the subsoil and the parent material.

EXERCISE C *Relationships between statements*

Part 1 *Exemplification and explanation*

Study the following sentences:

(i) When weathering processes are largely physical, the composition of the soil is very similar to that of the parent rock. In arid regions weathering is mostly by physical means.

(ii) The smaller the size of a particle, the greater is its reactivity. Smaller sized particles can react or combine with water, nutrients and humus more easily than larger sized particles.

The relationship between the sentences in (i) is one of *exemplification.* We can relate these sentences as follows:

EXAMPLE

When weathering processes are largely physical, the composition of the soil is very similar to that of the parent rock. { *For example* / *For instance* } in arid regions weathering is mostly by physical means.

OR When weathering processes are largely physical, the composition of the soil is very similar to that of the parent rock. In arid regions, { *for example,* / *for instance,* } weathering is mostly by physical means.

The relationship between the sentences in (ii) is one of *explanation.* We can relate these sentences as follows:

EXAMPLE

The smaller the size of a particle, the greater is its reactivity. { *That is to say,* / *In other words,* } smaller sized particles can react or combine with water, nutrients and humus more easily than larger sized particles.

Relate the sentences indicated with the following expressions as in the examples above. Write out both sentences. Replace and re-order the words in the sentences where necessary.

(a) Unit 2; 10+11: for example
(b) Unit 2; 20+21: (i) in other words (ii) that is to say

Part 2 *Consequence and contrast*

Relate the sentences indicated with the following expressions. Write out all the sentences. Replace and re-order the words in the sentences where necessary.

(a) 6+7: (i) however (ii) although
(b) 8, 9, 10+11: (i) as a result (ii) thus

(c) 21 + 22: although
(d) 26 + 27: (i) therefore (ii) hence

II LANGUAGE IN USE

EXERCISE A *Making tables from descriptions*

Read this description of a soil profile.

Profile of Soil A: Red Earth

The A horizon extends to a depth of 36 cm. The soil consists of a brownish red sandy loam. It has a porous and friable granular structure which is mixed with pebbles. The B horizon extends from 36 cm to 130 cm and is red in colour. It is a sandy loam, gravelly in structure with large quantities of pebbles. The C horizon, which extends down to 244 cm, has a yellowish white colour. It is sandy, with a structure which is a cemented and compact mass, made up of decomposed felspars.

Now study the following table which summarizes the information presented in the description above.

Profile of Soil A: Red Earth

Horizon	*Depth (cm)*	*Colour*	*Soil type (textural class)*	*Structure*	*Other features*
A	0–36	brownish red	sandy loam	porous and friable granular	mixed with pebbles
B	36–130	red	sandy loam	gravelly	mixed with large quantities of pebbles
C	130–244	yellowish white	sandy	cemented and compact mass	decomposed felspars

Read the following descriptions of soil profiles and make tables presenting the information as in the example above.

1. *Profile of Soil B: Laterite*

The A horizon extends down to 61 cm. Its colour is brownish deep red. The textural class of the soil is a clay loam. It has a loose granular structure which is mixed with gravel. The B horizon which extends from 61 cm to 259 cm is a clay loam. It is bright red with a gravelly laterite structure. The C horizon below 259 cm is red, mottled with yellow and whitish colours. It is rocky consisting of honey-combed laterite.

2. *Profile of Soil C: Mountain and hill soil*

The A horizon extends to a depth of 18 cm. Light brownish grey in colour, the soil type is a sandy loam with a friable granular structure. Another feature is that it is slightly sticky when wet. The B horizon extends from 18 cm to 66 cm. It is yellowish brown in colour and consists of a moist clay loam having a blocky, very hard and compact structure. It is also sticky when wet. The C_1 horizon goes down to 97 cm and is yellowish brown in colour. A moist clay loam, the soil has a blocky, extremely firm structure which is difficult to cut. It is very sticky when wet, and some parent material is mixed with the soil. Below 97 cm, the C_2 horizon is yellowish brown, a clay loam which is soft and easy to cut when moist. It has a laminated and compact clay shale structure.

EXERCISE B *Writing descriptions from tables*

Part 1

There are various patterns for describing the structure of the soil.

EXAMPLES

The soil is a loam *that is* slightly compact *in structure.*
= The soil is a loam *with a* slightly compact *structure.*
= The soil is a loam *with a structure that is* slightly compact.
= The soil is a loam *that has* a slightly compact structure.
= The soil is a loam *having a* slightly compact structure.

Rephrase the sentences below in each of the other ways shown in the examples above.

(a) The soil is a sandy loam that is granular in structure.
(b) The soil is a sandy loam with a cemented and compact structure.
(c) The soil is a silty loam with a structure that is blocky and weakly granular.
(d) The soil is a clay loam that has a laminated and compact structure.

Part 2

Note the following kinds of expressions used in the descriptions of soil profiles in Exercise A above:

(a) Descriptive labels for the textural class of different soils. The following table summarizes the possible combinations:

ADJECTIVE	NOUN
sandy silty	clay clay loam
gravelly silt(y) sandy clay(ey)	loam
loamy gravelly	sand

(b) Expressions used to describe the colour of the soil:

(i) using *-ish*: brownish grey, yellowish
(ii) comparing the colour with some other object: ash grey
(iii) using adjectives such as *dark, light, deep* etc.:
bright red, brownish deep red.

Study the following table and then complete the paragraph below describing soil profile D.

Profile of Soil D: Black soil

Horizon	*Depth (cm)*	*Colour*	*Soil type (textural class)*	*Structure*	*Other features*
A_1	0–15	brownish grey	coarse sandy clay loam	crumb	contains pieces of quartz
A_2	15–61	dark grey	heavy clay loam	hard crumb	
B	61–137	brownish or whitish grey	heavy clay loam	hard crumb	contains white and dark carbonate nodules
C	below 137	ash grey	heavy clay loam		mixed with disintegrating rock

...... extends to a depth of .... in colour, the soil type is ...... with a ...... structure. A feature is the presence of .... extends from ...... and is ...... in colour. It consists of a ...... having a ...... The B horizon goes down to ...... and is ...... in colour. The textural class of the soil is a ...... The soil has a ...... which contains ...... Below 137 cm the C ...... is ......, a ......

Now write descriptive paragraphs of your own based on the tables below.

1. *Profile of Soil E: Desert soil*

Horizon	*Depth (cm)*	*Colour*	*Soil type (textural class)*	*Structure*	*Other features*
A	0–18	pink	silty loam	blocky, tending to be weakly granular	dry and hard
B	18–58	light brown	silty clay loam	weakly laminar, developing to blocky and nutty	somewhat porous and dry
C	58–127	pinkish to slight brown	silty clay loam	compact and hard	somewhat gritty

2. *Profile of Soil F: Alluvial soil*

Horizon	*Depth (cm)*	*Colour*	*Soil type (textural class)*	*Structure*	*Other features*
A	0–15	ash grey	clayey loam	compact and cloddy	alkaline and calcareous
B	15–107	dark grey	clayey loam	compact and hard, also very cloddy	alkaline with nodules
C	107–182	grey	loam	compact	alkaline with an increasing number of lime nodules.

III GRAMMAR

EXERCISE A *Comparative sentences*

We can make comparative statements using the following patterns:

(i) Particles of fine sand are *coarser than* particles of clay.
= Particles of clay are *not as coarse as* particles of fine sand.

(ii) Transported soils are *more common than* sedentary soils in humid regions.
= Sedentary soils are *less common than* transported soils in humid regions.

The sentences in (i) use the *comparative form of the adjective + than*, and *not as + adjective + as*. The sentences in (ii) use *more* and *less + adjective + than*. Such adjectives are usually of two syllables or more. Note that two syllable adjectives ending in *y* have comparative forms ending in *-ier*. For example: heav*y* – heav*ier*: eas*y* – eas*ier* etc.
Compare the information presented in the tables and diagram below using the adjectives given. Use *ADJ-er + than* and *not as ADJ as* constructions to express the comparison.

1. TABLE 1

Soil particles	*Diameter (mm)*
gravel	2·0 or more
coarse sand	2·0–0·2
fine sand	0·2–0·02
silt	0·02–0·002
clay	0·002 or less

Compare the particle size of
(a) fine sand with silt (fine)
(b) clay with fine sand (coarse)
(c) fine sand with gravel (coarse)

2. TABLE 2

4	5	6	7	8	9
strongly acid	*moderately acid*		*neutral*	*moderately alkaline*	*very alkaline*
most soils in humid regions when not limed					occur only in arid regions

The pH scale

Compare these soils in terms of *more* or *less* acidity or alkalinity.
(a) A soil with a pH value of 4·0 with a soil with a pH value of 6·5
(b) A soil with a pH value of 8·0 with a soil with a pH value of 9·8

3.

TABLE 3

A: light sandy soil		*B: heavy clay soil*
	%	%
gravel	1·2	1·3
sand	37·4	33·7
silt	38·6	26·9
clay	11·8	28·3
humus	4·5	7·8
other contents	6·5	2·0

Compare the contents of these soils in terms of

(a) their sand content
(b) their silt content
(c) their clay content
(d) their humus content

Begin the statements as follows:

(i) The percentage of sand/silt etc. in a light sandy soil ......
(ii) The sand/silt etc. content of a heavy clay soil ......
(iii) A light sandy soil has......

4.

TABLE 4

	TOP SOIL	SUB-SOIL
colour	dark	light
particle sizes	coarse	fine
living organisms	many	few
elements for plant food	rich	poor

Using the information in this table compare the characteristics of the top soil and the sub-soil in terms of

(a) colour: (i) The top soil is ......
(ii) The sub-soil is ......
(b) particle sizes: (i) The particle sizes of the top soil ......
(ii) The particle sizes of the sub-soil ......
(c) living organisms: (i) There are ...... in the top soil ......
(ii) There are ...... in the sub-soil ......
(d) elements for plant food: (i) The top soil is ......
(ii) The sub-soil is ......

EXERCISE B *Contrastive sentences*

Study the following sentences about information contained in Table 1 in Exercise A above.

> Particles of coarse sand are between 2·0 and 0·2 mm in diameter, *whereas* particles of fine sand are between 0·2 and 0·02 mm in diameter.
>
> = The diameter of particles of coarse sand is between 2·0 and 0·2 mm, *whereas* the diameter of particles of fine sand is between 0·2 and 0·02 mm.

Make contrastive statements about the information presented in the tables in Exercise A using *whereas* to connect the statements together as in the example above.

1. Table 1: particles of silt and particles of clay
   (a) Particles of silt ......
   (b) The diameter of ......

2. Table 2: the pH value of soils in humid regions and the pH value of soils in arid regions.
   (a) The pH value of ......
   (b) Soils in humid ......

3. Table 2: the pH value of moderately acid soils and the pH value of strongly acid soils.
   (a) The pH value of ......
   (b) Moderately ......

4. Table 3: the clay content of a light sandy soil and a heavy clay soil.
   (a) The clay content of ......
   (b) A light sandy soil ......

5. Table 4: elements for plant food in the top soil and in the sub-soil.
   (a) The top soil ......
   (b) The sub-soil ......

EXERCISE C *Making comparisons by inference*

Part 1

Complete these sentences with the *comparative form* of an appropriate *adjective*, or *more/less*.

(a) Soils show great variations in their sizes and arrangements of their constituent particles. A sandy soil has larger particles than a clay soil. A sandy loam has _____ particles than a clay loam.

(b) Soils also vary greatly in colour. A brightly coloured soil indicates a higher degree of oxidation. So, a red soil has been _____ oxidized than a black soil.

(c) As agriculture becomes more intensive, the soil may be modified by those who form it. A soil can be made less alkaline by adding sulphur. Any soil can be made _____ acid by adding lime.

(d) The sulphur content of certain soils in Oregon, U.S.A., is less than 0·15%. On average in the U.S.A. soil contains several times this amount. We may say then that most soils contain _____ sulphur than those of Oregon.

(e) All the spaces, or pores, in between the solid soil particles are filled with air and water. The proportions of air and water which are contained in these pores are determined mainly by the size of the pores. The bigger the size of the pores the more air and _____ water the soil will contain. Thus, sandy soils contain larger pore spaces, but the total amount of pore space is _____ because the particles are _____. On the other hand, clay soils contain smaller pore spaces but the total amount of pore space is _____ because the particles are much _____. As a result, clay soils are generally _____ than sandy soils.

Part 2

Study the following short passages and complete the statements about them below.

(a) Granites and other igneous rocks are usually divided into acidic, intermediate and basic rocks depending on their silica content. This may vary between 40% for basalts, which are dark, glassy basic rocks, to twice that percentage in granites.

Write *less* or *more* in the spaces provided.

(i) A granite rock contains _____ silica than a basic rock.
(ii) There is _____ silica in basalt than in granite.
(iii) Acid rocks contain _____ silica than basic rocks.
(iv) Basalt is not an acidic rock because it contains _____ than 75% silica.
(v) Granite is very acidic because it contains _____ than 75% silica.

(b) The colours of soils are closely related to their condition of aeration. In well-drained soils iron compounds are oxidized to their ferric state, which is indicated by reds, yellows and browns. When good drainage is absent, soils tend to be grey, often with greenish grey or mottled sub-soils.

Write *less* or *more* or *better* or *poorer* in the spaces provided.

(i) A red soil is _____ drained than a grey soil.
(ii) A grey soil has been _____ oxidized than a yellow soil.
(iii) Greenish grey soils are _____ well drained than brown soils.
(iv) A red soil has _____ conditions of aeration than a grey soil.
(v) Well-drained soils are _____ aerated than badly drained soils.

(vi) Grey and greenish soils have _____ drainage than oxidized red soils.
(vii) There is _____ air in a well-drained soil than in a badly drained soil.

(c) The soil microbes which decompose organic materials grow best at pH 6·5. Near this pH, conditions are best for the availability of most plant nutrients. As the acidity increases, the availability of nearly all important nutrients diminishes. Phosphorous, in particular, is held as insoluble compounds in highly acid soils. As acidity decreases (i.e. as pH rises) iron, manganese, copper and zinc grow scarce. Most upland soils developed under forests in humid regions are too acid for the best growth of pasture grasses, vegetables and many other plants.

Write *less* or *fewer* or *more* or *not as good* or *higher* in the space provided.

(i) At a pH value of below 4 there are _____ important nutrients in the soil.
(ii) There is _____ soluble phosphorous in highly acid soils.
(iii) Iron, manganese, copper and zinc are _____ available in _____ alkaline soils.
(iv) Vegetables prefer conditions of _____ acidity than is found in most upland soils.
(v) Conditions are _____ for the decomposition of organic materials by soil microbes at low pH soil values.

IV SUMMARY CHECK

Complete the following text by filling in the blank spaces. Some of the expressions you will require are given below. A dotted line . . . . . . requires a phrase to be added and a straight line _____ requires a word.

is derived from	mixture	property
composition	composed of	sizes
non-solid	therefore	solids
chemical decomposition	sub-soil	smaller particles
		organic matter

The soil system is made up of mineral particles which are mixed with decomposed . . . . . . The top soil consists of this _____, which is so vital for plant growth. Below the top soil is the _____ which is largely . . . . . . mineral matter. In addition to the mineral and organic matter, called the soil _____, there are spaces between the . . . . . . which are taken up by . . . . . . to make up the _____ part of the soil. The accompanying diagram shows the volume _____ of a typical top soil. Amounts are approximate as the percentage of certain constituents e.g. . . . . . ., is constantly varying.

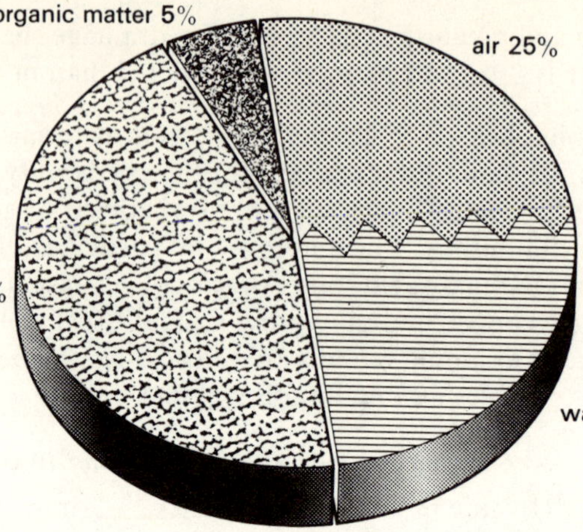

Soil solids consist mainly of particles of various _____. All particles between 0·002 mm and 0·02 mm are _____. Particles . . . . . . 0·02 mm are sand particles, coarse sand and . . . . . . Particles smaller than silt are . . . . . . Clay is able to absorb a great deal of_____ owing to the amount of pore space between . . . . . . Sand does not have this _____. _____, a soil which contains more clay is able to hold more _____ than a _____with . . . . . .

The mineral material in all soil . . . . . . parent material by the process of _____ which breaks down rocks into . . . . . . by mechanical disintegration and . . . . . .

4 Drainage and Irrigation

I READING AND COMPREHENSION

[1]One meaning of drainage is the natural ability of the soil to allow a downward movement of water. [2]The ease with which water can pass through a soil depends on the proportions in it of coarse and fine particles such as sand and clay. [3]The finer the particles become, the more slowly the water percolates, or passes, through the soil. [4]So heavy soils such as clay are more impermeable than light soils.

[5]When there is too much water in the soil, some of it must be drained off. [6]This is the other meaning of drainage: the removal of excess water from the soil by ditching or tiling. [7]This is done in order to maintain a correct balance of air and water in the soil. [8]Good drainage makes a soil easier to work. [9]It also helps to increase the feeding area of the soil for the roots of plants. [10]Another advantage is that a well-drained soil will have enough air for aerobic bacteria to break down humus and so provide food for the plant.

(a) Water can pass through a permeable soil more easily than through an impermeable soil.
(b) Water percolates through a soil more slowly when the soil particles are coarse.
(c) A correct balance of air and water is maintained by drainage.

[11]Ditching is one of the most important techniques for draining land. [12]Ditches can be cut at certain intervals between the crops. [13]These will remove surface water. [14]They should be wide and straight, with sloping sides, and they should be regularly cleaned. [15]Another important technique is tile drainage. [16]Porous drainage tiles may be laid in or on the land and these will help to draw off the surplus water. [17]The distance between the drains will depend on the level of the land, the permeability of the soil, and the amount of rainfall. [18]For very heavy soils mole drainage can be used. [19]This technique is used where water accumulates underground. [20]A tunnel is bored about 3 inches in diameter through the earth at a depth of about 2 feet.

(d) The most important technique for draining land is ditching.
(e) If the soil is impermeable tile drains should be laid far apart.
(f) Underground water can be drawn off by using mole drains.

[21]Where and when water is in short supply, irrigation is needed to make up the deficit. [22]We should distinguish between the collection of water and its application. [23]There are two main sources of irrigation water: surface water and ground water. [24]The former may be obtained from rivers, lakes or reservoirs, and the latter is provided by underground water deposits. [25]Irrigation from rivers is mainly along canals from dams which have been built across the rivers. [26]The water collects behind the dam during the wet season. And it is applied in the fields later during the dry season. [27]Subterranean water is obtained by digging or drilling a well. [28]In either case it is necessary to lift the water before it can be used for irrigation.

(g) Rivers, lakes or reservoirs provide surface water for irrigation.
(h) Irrigation takes place during the wet season.
(i) Before subterranean water can be used it must be lifted from a well.

[29]The amount of water which is required for irrigation depends on a number of factors. [30]It depends, firstly, on the type of soil, and the deficit in the soil. [31]By this we mean the amount of water which is needed to bring the soil to full capacity. [32]It also depends on the type of crop, the stage of growth of the crop and the amount which it will use at that particular time. [33]The irrigation requirement of a crop is not the same throughout its growing period. [34]Most plants require larger quantities of water during the later stages than in the earlier stages. [35]Sugar cane needs heavier irrigation or more frequent irrigation from about the sixth or seventh month onwards. [36]In the same way, grain crops require their maximum irrigation during the time earheads are forming.

Solutions

(a) As the particles in a soil become finer, water passes through the soil more _____. (3)
e.g. Water passes more slowly through heavy _____ soils such as clay. (4)
i.e. It is more difficult for ...... through an impermeable soil.
∴ It is _____ for water to pass through a _____ soil.
i.e. *Water can pass through a permeable soil more easily than through an impermeable soil.*

(b) As the particles in a soil become _____ water passes through the soil more slowly. (3)
∴ The coarser the particles become, the more _____ the water _____ through the soil.
i.e. Water passes more quickly through the soil when the soil particles ......
= Water percolates (= passes) through a soil more quickly (NOT more slowly) when the soil particles are coarse.

(c) The removal of excess water from the soil is known as _____. (6)
Soils are drained so as to maintain . . . . . . (7)
i.e. *A correct balance of air and water is maintained by drainage.*

(d) In order to drain land, ditching is . . . . . . techniques. (11)
ONE OF THE MOST important techniques ≠ THE _____ important technique i.e. there are others e.g. _____ drainage. (15)
∴ It is NOT TRUE to say that the most important technique for draining land is ditching.

(e) The distance between tile drains depends partly on whether the soil is permeable or _____. (4)
If the soil is impermeable the water will pass . . . . . . (see solution **a**)
If the water passes slowly through the soil MORE _____ will be needed. (17)
∴ Tile drains should NOT be laid . . . . . .
i.e. Tile drains should NOT be laid far apart if the soil is _____.
= If the soil is impermeable tile drains should NOT be laid far apart.

(f) The technique of mole drainage can be used to draw off water which . . . . . . (19)
Water which accumulates underground, i.e. _____ water.
∴ Underground water can be . . . . . . by using _____ drains.
Underground water can be drawn off by using mole drains.

(g) The sources of water used for irrigation are . . . . . . and ground water. (23)
The former (i.e. surface water) may be obtained from . . . . . . (24)
i.e. Rivers, lakes and reservoirs provide . . . . . . for _____.
Rivers, lakes and reservoirs provide surface water for irrigation.

(h) During the wet season water _____ behind the dam. (26)
i.e. Collection is one stage of _____.
During the dry season water is _____ to the fields. (26)
i.e. _____ is another stage of irrigation.
∴ Irrigation takes place BOTH during the _____ season (_____ stage) AND the _____ season (_____ stage).
Irrigation takes place BOTH during the wet season (collection stage) AND during the dry season (application stage).

(i) In order to obtain subterranean water a well must be _____ or _____. (27)
In either case, whether a well is dug or drilled, it is necessary . . . . . . (28)
i.e. Before . . . . . . can be used for _____ it is necessary to lift the water from a _____.
it is necessary to lift the water i.e. the water _____ be lifted.
∴ *Before subterranean water can be used for irrigation it must be lifted from a well.*

EXERCISE A *Contextual reference*

1. In sentence 2 *it* refers to:
   (a) water
   (b) a soil
2. In sentence 5 *it* refers to:
   (a) water
   (b) the soil
3. In sentence 7 *this* refers to:
   (a) excess water in the soil
   (b) the removal of excess water from the soil
4. In sentence 13 *these* refers to:
   (a) the crops
   (b) intervals
   (c) ditches
5. In sentence 14 *they* refers to:
   (a) the crops
   (b) intervals
   (c) ditches
6. In sentence 19 *this technique* refers to:
   (a) ditching
   (b) tile drainage
   (c) mole drainage
7. In sentence 24 *the latter* refers to:
   (a) ground water
   (b) surface water
   (c) rivers, lakes or reservoirs
8. In sentence 26 *it* refers to:
   (a) the wet season
   (b) the dam
   (c) the water
9. In sentence 31 *this* refers to:
   (a) the deficit in the soil
   (b) the type of soil
10. In sentence 32 *it* refers to:
   (a) full capacity
   (b) the amount of water which is needed
   (c) the soil

EXERCISE B *Rephrasing*

Rewrite the following sentences replacing the words printed in italics with expressions from the text which have the same meaning.

1. Water *passes through the soil* quickly when the *ease with which water can pass through the soil* is high.

2. *Removing excess water from the soil* helps to increase the feeding area of the soil for the plant roots.
3. *Laying porous drainage tiles on the land* helps to remove surplus surface water.
4. *Water from rivers, lakes or reservoirs* may have to be transported along canals a long distance to the fields, whereas *water from underground deposits* lifted from a well may be close to the crops.
5. How much water a crop requires from irrigation depends partly on *the shortage of water* in the soil.

EXERCISE C *Relationships between statements*

Part 1 *Addition: reinforcement and similarity*

Study the following sentences:

(i) Good drainage makes a soil easier to work. It helps to increase the feeding area of the soil for the roots of plants.

(ii) Sugar cane needs heavier irrigation from about the sixth or seventh month onwards. Grain crops require the maximum irrigation during the time earheads are forming.

The relationship between the sentences in (i) is one of *addition, reinforcing* what has been said previously i.e. the advantages of good drainage. We can relate these sentences as follows:

EXAMPLE

Good drainage makes a soil easier to work.
In addition, / *Moreover,* / *Furthermore,* } it helps to increase the feeding area of the soil for the roots of plants.

OR Good drainage makes a soil easier to work. It *also* helps to increase the feeding area of the soil for the roots of plants.

The relationship between the sentences in (ii) is also one of *addition*, adding a statement which is *similar* to what has been said previously i.e. crops need more irrigation during the later stages of growth. We can relate these sentences as follows:

EXAMPLE

Sugar cane needs heavier irrigation from about the sixth or seventh month onwards. *Similarly,* / *In the same way,* } grain crops require the maximum irrigation during the time earheads are forming.

Relate the sentences indicated with the following expressions as in the examples above. Write out all the sentences. Replace and re-order the words in the sentences where necessary.

(a) 9+10: (i) in addition (ii) furthermore
(b) 30, 31+32: (i) moreover (ii) furthermore
(c) Unit 1; 7, 8+9: in addition
(d) Unit 1; 18+19: similarly
(e) Unit 2; 13, 14+15: (i) furthermore (ii) in addition
(f) Unit 3; 11+12: also
(g) Unit 3; 30+31: (i) moreover (ii) also

Part 2 *Consequence, contrast, exemplification and explanation*

Relate the sentences indicated with the following expressions. Write out all the sentences. Replace and re-order the words in the sentences where necessary.
(a) 3+4: (i) therefore (ii) hence
(b) 12+13: with the result that
(c) 30+31: (i) that is to say (ii) or
(d) 32+33: this is because
(e) 34+35+36: (i) for example (ii) similarly

II LANGUAGE IN USE

EXERCISE A *Classification and definition*

Refer to the first part of Reading Passage of Unit 3, and then study the following diagram.

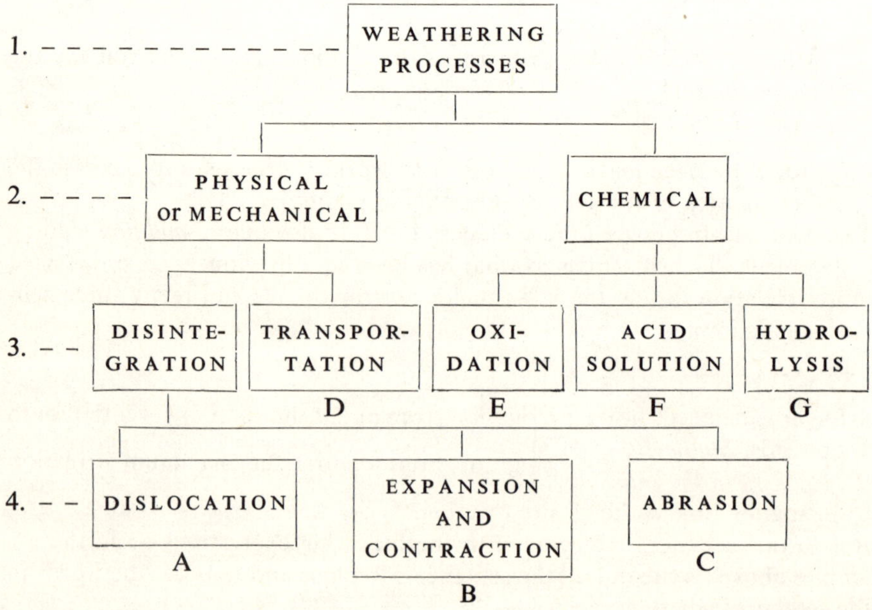

The diagram is a *classification* of weathering processes. There are 4 levels of *generalization*. Items at a higher level are more general than items at a lower level. Items at a lower level are *kinds* of items at a higher level.

Part 1

Arrange the following weathering agents under the correct weathering process or processes labelled A to G in the diagram. Make statements about each agent. For example:

(a) ice is A and C.
i.e. Ice is an agent of dislocation and abrasion.

(a) ice	(d) water	(g) wind
(b) heat	(e) carbonic acid	(h) rivers
(c) plant roots	(f) oxygen	(i) lichens and mosses
		(j) glaciers

Part 2

Write simple definitions about the following weathering processes.

EXAMPLE

Dislocation is a physical, or mechanical, weathering process.

(a) expansion and contraction	(d) oxidation
(b) abrasion	(e) acid solution
(c) transportation	(f) hydrolysis

Part 3

We may combine the definitions in Parts 1 and 2 as follows:

Dislocation is a physical weathering process. Agents of dislocation include water and ice.

OR Dislocation is a mechanical weathering process. Water, plant roots, ice and wind are agents of this process.

Write statements about the other processes listed in Part 2, above.
Where there is only one agent, write

The agent of x is y, or *y is the agent of x,*

where x = the process, and y = the agent.

EXERCISE B: *Definition, description and identification*

Part 1

Study the following table and the examples below it.

A *Name of process*	B *Agent(s)*	C *Mode of operation*
(a) Dislocation	water and ice	separate solid particles from the parent rock
(b) Expansion and contraction	heat	sets up tensions in rocks which break them apart
(c) Abrasion	wind and ice	grind away the surfaces of rocks
(d) Transportation	water	washes away solid particles and deposits them elsewhere as soil

A+C:	definition:	Dislocation is a physical weathering process whereby solid particles *are separated* from the parent rock.
B+C:	description:	Water and ice act as physical weathering agents *by separating* solid particles from the parent rock.
B+C+A:	description+ definition:	Water and ice act as physical weathering agents by separating solid particles from the parent rock. *This process is known as* dislocation.
C+A+B:	description+ identification:	Solid particles are separated from the parent rock *by* the mechanical process of dislocation. Water and ice *are* (*the chief*) *physical agents of* this weathering process.
B+A+C:	identification+ description	Water and ice are (the chief) agents of the physical weathering process of dislocation. *In this process* solid particles are separated from the parent rock.

Write definitions, descriptions and identifications for (b), (c) and (d), as in the examples above.

Part 2

Write definitions, descriptions and identifications about these *chemical weathering processes* and *agents* as in Part 1 above.

A *Name of process*	B *Agent*	C *Mode of operation*
(a) Acid solution, or dissociation	carbonic acid	dissolves mineral substances in a weak acid solution
(b) Oxidation	oxygen	combines certain elements or compounds with oxygen to form oxides
(c) Hydrolysis	water	exchanges constituent parts of water and minerals in a reaction

EXERCISE C: *Classifications in diagrams and paragraphs*

Part 1

Refer to your textbooks if necessary and classify the following items in the form of diagrams as in Exercise A above.

(a) rocks – igneous – sedimentary – metamorphic – acidic – basic stratified (water laid) – non-stratified (wind or ice laid) – granites, gneisses – basalt – limestone – loess – marble, slate
(b) sources of irrigation water – surface water – ground water – flowing water – still water – hand dug shallow wells – tube-bored deep wells – rivers – lakes, reservoirs
(c) plants – ephemeral – annual – biennial – perennial – herbacious perennials – woody perennials – weed groundsell – peas – cabbages, sugar beet – lucerne, rhubarb – trees, shrubs
(d) drainage systems – ditches – tile drainage – mole drainage – natural tile system – parallel tile system – grid pattern – herring-bone pattern
(e) soils – soils of humid regions (pedalfers) – soils of arid or semi-arid regions (pedocals) – soils of tropical zones – soils of temperate zones – soils of arctic zones – soils where rainfall 12–25″ per year – soils where rainfall 10–15″ per year – soils where rainfall less than 10″ per year – latosols – podzols – tundra – chernozems – chestnut-brown soils – desert and saline
(f) inorganic fertilizers – nitrogenous – phosphatic – potash – sulphate of ammonia – calcium cyanamide – nitrate of soda – nitro-chalk and nitrashell – superphosphate – triple superphosphate – rock phosphates – muriate of potash – sulphate of potash
(g) sources of plant nutrients — from the air – from the soil – from soil water – from mineral substances – primary nutrients – secondary nutrients – micronutrients – oxygen, carbon and nitrogen – hydrogen – nitrogen, phosphorous and potassium – calcium, magnesium and sulphur – iron, copper and zinc

Part 2

Study the following tables carefully.

(a)

Weathering processes	can be divided into two	classes: groups: kinds:	physical weathering and chemical weathering.
	may be	divided into classified as	physical and chemical.

(b)

Two	groups classes types kinds	of	weathering process	can be may be	distinguished:	physical and chemical.
There are two		of weathering process:				

(c)

Examples of physical	weathering agents	include include agents such as	ice, heat wind and water.
Physical		are of various types including	

1. Complete the paragraph below using information from the diagram in Exercise A on page 48. Note that many of the sentences are similar in construction to the tables above.
   Weathering processes can be classified as ...... and ...... The former may be divided into processes of ...... and ......, while ...... processes may be divided into those of ......,...... and ...... There are three kinds of processes of disintegration ......: ......, ... and ...... Agents of these processes include ......, ..., and ...... Examples of ...... agents include agents such as ......, ... and ......

2. Using the patterns given in the tables above, write out classifications of items you have arranged in diagrams in Part 1, on page 51.

EXAMPLES

(i) Rocks can be divided into three main classes: igneous, sedimentary and metamorphic. There are two types of igneous rock, acidic and basic.

Acidic igneous rocks include rocks such as granites and gneisses while basalt is an example of a basic rock. In addition, two kinds of sedimentary rock can be distinguished: stratified rocks, or those which have been water laid, such as limestone, and non-stratified rocks or those which have been wind or ice laid, such as loess. Examples of metamorphic rocks include marble and slate.

OR

(ii) Rocks may be classified as igneous, sedimentary and metamorphic. Two kinds of igneous rock can be distinguished: acidic rocks, such as granites and gneisses, and basic rocks, such as basalt. In addition, sedimentary rocks can be divided into stratified sedimentary rocks, or those which have been water laid – for example, limestone – and non-stratified sedimentary rocks, or those which have been wind or ice laid – loess, for example. Metamorphic rocks include rocks such as marble and slate.

EXERCISE D: *Classification according to defining characteristics.*

Classifications are made according to some principle of classification, and statements are made in the active or passive form.

EXAMPLE

1. (i) We may (or: can) classify soils according to the climatic factors in their development.
   (ii) Soils may (or: can) be classified according to the climatic factors in their development.
2. (i) We may (or: can) classify plants according to how long they live.
   (ii) Plants may (or: can) be classified according to how long they live.

Such statements may be followed by examples of particular types or groups and a statement of their defining characteristics.

EXAMPLE

Object	*Principle of classification*	*Examples*	*Characteristics*
Soils	climatic factors in their development	pedalfers	develop in humid regions
		pedocals	develop in arid and semi-arid regions

= Soils may be classified according to the climatic factors in their development. For example, pedalfers develop in humid regions, *but* pedocals develop in arid and semi-arid regions.

Make statements about how the following objects may be classified as in the example above. You will need to add the following kinds of words, where appropriate:

(a) definite and indefinite articles;
(b) pronouns and adjectives like *they* and *their*;
(c) forms of the verb *be*.

A	B	C	D
Object	*Principle of classification*	*Examples*	*Characteristics*
1. Plants	how long they live	annual plants	complete life cycle in single season
		biennial plants	require two years to complete life cycle
2. Pedalfers	temperature of climatic zone in which occur	latosols	found in tropical zones
		podzols	occur in temperate zones
3. Plant nutrients	source	oxygen and carbon	derived from air
		minerals such as phosphorous, potassium, calcium, iron and zinc	extracted from mineral substances in soil
4. Soils	proportions of sand, silt and clay each contain	light sandy soil	contains low proportion of clay, and higher proportion of sand and silt
		heavy clay soil	contains almost equal proportions of sand, silt and clay
5. Soils	texture	sandy soil	coarse to touch and particles do not stick together
		clay soil	tough and plastic

A	B	C	D
Object	*Principle of classification*	*Examples*	*Characteristics*
6. Crops	use made of them	cereal or grain crops, such as wheat and barley	grown primarily for seed which is used for food
		forage and pasture crops, such as legumes	grown primarily for leafy parts which are used for livestock feed
7. Weathering agents	natural force involved	physical agents such as water, ice and wind	change size, form and shape of rock and mineral debris
		chemical agents such as oxygen and natural acids	alter chemical composition of rocks and minerals
8. Soil organisms	oxygen requirements	anaerobic bacteria	do not require oxygen in order to live
		aerobic bacteria	live on free oxygen in the air

III GRAMMAR

Part 1

We can express the purpose for which something is done by using a *to-infinitive*.

EXAMPLES

(a) We cut ditches at certain intervals between the crops. *Our purpose* (*aim*) *in doing this* is to remove surface water.

= (i) *We cut* ditches at certain intervals between the crops *to remove* surface water.

= (ii) Ditches *are cut* at certain intervals between the crops *to remove* surface water.

(b) We may lay porous drainage tiles in or on the land. *Our object* (*aim*) *in doing this* is to draw off the surplus water.

= (i) We may lay porous drainage tiles in or on the land *to draw off* the surplus water.

= (ii) Porous drainage tiles *may be laid* in or on the land *to draw off* the surplus water.

Rewrite the following sentences as in (i) and (ii) above.

(a) A plant uses sunlight. The purpose in using it is to combine carbon dioxide with water to form carbohydrates.
  (i) A plant uses sunlight ...... carbon dioxide with water to form carbohydrates.
  (ii) Sunlight ...... by a plant ...... carbon dioxide with water to form carbohydrates.

(b) In countries with inadequate rainfall we irrigate the land. Our object in doing this is to supply enough moisture for satisfactory plant growth.
  (i) In countries with inadequate rainfall we irrigate the land ...... enough moisture for satisfactory plant growth.
  (ii) In countries with inadequate rainfall the land .... enough moisture for satisfactory plant growth.

(c) The roots of plants such as sugar beet, carrots and parsnips store food. The object of doing this is to enable them to live more than one season.
  (i) The roots of plants such as sugar beet, carrots and parsnips store food ...... them to live more than one season.
  (ii) Food ...... by the roots of plants such as sugar beet, carrots and parsnips ...... them to live more than one season.

(d) We should add fertilizers or manures to the soil. Our aim in doing this is to encourage plant growth.
  (i) We should add fertilizers or manures to the soil ...... plant growth.
  (ii) Fertilizers or manures ...... to the soil ...... plant growth.

(e) If a soil is very acid we can spread lime on it. Our purpose in doing this is to correct the acidity.
  (i) If a soil is very acid we can spread lime on it ...... the acidity.
  (ii) If a soil is very acid lime ...... on it ...... the acidity.
  (iii) Lime ...... on a soil which is very acid ...... the acidity.

Part 2

Instead of a *to-infinitive* we may use *in order to + infinitive* or *so as to + infinitive* to express purpose.

EXAMPLE

We need irrigation. The reason for this is to make up for lack of natural rainfall.

= (i) Irrigation is needed *in order to* make up for lack of natural rainfall.
= (ii) Irrigation is needed *so as to* make up for lack of natural rainfall.

Join the following pairs of sentences together using the passive forms of the verb followed by (i) *in order to* and (ii) *so as to* as in the examples above.

(a) Tile drains may be laid in or on the land. The reason for this is to remove surplus water from the fields.
(b) We apply inorganic materials to the soil. Our object in doing this is to increase the supply of one or more of the essential nutrients such as nitrogen, phosphorous, and potash.
(c) We add bulky materials to the soil like animal or green manure. Our aim in doing this is to improve the physical condition of the soil and to keep up its humus content.
(d) We should keep ditches as straight as possible and on an even gradient. The reason for this is to prevent them from silting up.
(e) A plant transforms water and carbon dioxide into sugar and starch in the presence of light. Its purpose in doing this is to provide the plant with food and energy for growth and reproduction.

IV SUMMARY CHECK

Complete the following text by filling in the blank spaces. Some of the expressions you will require are given below. A dotted line ...... requires a phrase to be added, and a straight line _____ requires a word.

digging or drilling	provided	length
get rid of	obtained	drained
tube-bored	function	type of crop
must be		

Where and when water is in short supply _____ is needed in order to make up for natural rainfall. There are two main sources of _____ water: surface water and ...... Surface water is _____ by the flowing waters of rivers or is _____ from the still water of tanks, ponds or artificial _____. Subterranean water is tapped by ...... wells. Wells may be shallow wells which have been dug by hand or _____ wells which have been _____. In either case water ...... lifted before it can be used for _____.

For a given type of soil the amount of water which is required varies with the ...... – the plant, its physiological make-up and the _____ of the growing season.

Where and when there is too much water in the soil some of it must be _____ off. Thus, _____ and _____ are like two faces of a coin. While the purpose of irrigation is to arrange for sufficient _____ in the soil for satisfactory ......, the _____ of drainage is to remove _____ moisture from the root-zone. To avoid an excess of moisture, care must be taken to ...... any collection of water either above or below ground surface. There are three methods of doing _____: by ditching, by laying ...... and by using ....

5 Manures and Fertilizers

I READING AND COMPREHENSION

[1]Plant growth cannot continue if there is not a supply of minerals in a soil. [2]The materials which are available for this purpose can be divided into two groups: the bulky, organic materials which are called manures, and the more concentrated, inorganic chemical substances which are called fertilizers. [3]Farmyard manure, or dung, consists of a mixture of litter, solid excreta and urine. [4]It contains three most important substances for plant materials – nitrogen, phosphate and potash. [5]Manure is added to the soil for several reasons. [4]It improves the physical condition of the soil. [7]It also keeps up the level of humus in the soil, and maintains the best conditions for the activities of soil organisms. [8]Finally, it makes up for the plant nutrients which have been removed by crops or lost by leaching and soil erosion.

(a) Manures are as concentrated as fertilizers.
(b) Nitrogen, phosphate and potash are the three most important substances for plant materials.
(c) The level of humus in the soil can be kept up by adding manure.

[9]Another kind of manure is green manure. [10]This includes leguminous crops which grow quickly such as clover and lucerne. [11]Such crops supply additional nitrogen as well as organic matter. [12]A leguminous crop which is ploughed under will add as much nitrogen to the soil per acre as 3 to 10 tons of farmyard manure.

[13]Fertilizers are usually classified according to the particular food element which forms their main constituent. [14]So, they may be grouped as nitrogenous fertilizers, phosphatic fertilizers, potassic fertilizers and so on.

[15]The most commonly used fertilizer which contains nitrogen is ammonium sulphate, which is made from ammonia and sulphuric acid, and which contains 21% nitrogen. [16]This element encourages rapid vegetative growth and gives plants a healthy green colour. [17]Another valuable nitrogenous fertilizer is urea, which is made from ammonia and carbon dioxide, and contains 46% nitrogen.

(d) Nitrogen is supplied to the soil by quick-growing leguminous crops.
(e) Nitrogen is the food element which is the main constituent of a nitrogenous fertilizer.
(f) Ammonium sulphate is a nitrogenous fertilizer.

[18]The most widely used phosphatic fertilizer, superphosphate, is made by treating mineral phosphate with sulphuric acid. [19]Phosphorous stimulates the formation of a plant's roots, and promotes fruit and seed production. [20]Tropical soils are very often poor in this element.

[21]Finally, wherever high crop yields are expected, potash is used together with nitrogen and phosphorous. [22]Potassium makes the plant tissues stronger. [23]This helps the plant to withstand mechanical damage such as broken branches and torn leaves. [24]In this way the entry of disease bearing agents, or pathogens, such as bacteria and fungi, is prevented. [25]Potassium is important for all plants but particularly so for those that produce oil and starch or sugars. [26]Oil palm and tapioca plants require potassium in large amounts. [27]It is usually supplied in the form of muriate of potash (potassium chloride), which contains 50 to 60% potassium oxide (K_2O) and sulphate of potash (potassium sulphate).

(g) Fruit and seeds will not be produced if a plant is not given a phosphatic fertilizer.
(h) Help in withstanding mechanical damage prevents the entry of pathogens into plants.

[28]All plants are affected by the degree of acidity or alkalinity of the soil. [29]The less the nutrient supply, the more acid the soil becomes. [30]Because mineral salts are basic, an acid soil has a low base content. [31]Acidity makes some elements unavailable to plants. [32]If a soil is very acid, with a pH value of less than 5·0, lime can be added to correct this acidity. [33]The main constituent of lime is calcium, an important plant food. [34]The presence of lime helps to make essential elements of plant food more easily available to plants. [35]Nitrogen, phosphorous and potassium are more easily available in a well-limed soil than in an acid soil.

Solutions

(a) Two groups of materials are being compared: _____ and _____. (2)
Manures are organic materials which are _____.
Fertilizers are inorganic chemical substances which are ......
i.e. Fertilizers are MORE concentrated than _____.
Fertilizers are MORE concentrated than manures. = Manures are NOT ...... fertilizers. (see Unit 3, Grammar, Exercise A)
Manures are NOT as concentrated as fertilizers.

(b) Nitrogen, phosphate and potash are substances that are found in ...... (3)
Farmyard manure contains these three ...... substances. (4)
Three most important substances = three *very important* substances.

≠ The three most important substances.
∴ It is NOT TRUE to say that nitrogen, phosphate and potash are THE three most important substances for plant materials.

(c) There are several reasons why _____ is added to the soil. (5)
One of the reasons is that manure keeps up ...... (7)
i.e. One of the consequences of adding manure to the soil is that the level of humus in the soil ......
i.e. By ...... the level of humus in the soil can be kept up.
= *The level of humus in the soil can be kept up by adding manure.*

(d) Leguminous crops such as clover and lucerne grow _____. (10)
i.e. Clover and lucerne are quick-growing ......
Leguminous crops supply the soil with ...... and ...... (11)
∴ Nitrogen is _____ to the soil by ......
Nitrogen is supplied to the soil by quick-growing leguminous crops.

(e) Fertilizers may be grouped as ...... (14)
Each of these fertilizers has a particular food element ...... (13)
The food element which forms the main constituent of a phosphatic fertilizer is phosphorous.
∴ The ...... which is ...... of an nitrogenous fertilizer is _____.
= *Nitrogen is the food element which is the main constituent of a nitrogenous fertilizer.*

(f) Ammonium sulphate is the most commonly used fertilizer ...... (15)
Nitrogen is the food element which is the main constituent of a ...... (See solution **e** above).
∴ Ammonium sulphate is ......
Ammonium sulphate is a nitrogenous fertilizer.

(g) The main constituent of a phosphatic fertilizer is _____. (see solution **e**)
Phosphorous _____ fruit and seed production. (19)
i.e. Phosphorous promotes, or stimulates, plants to produce ......
but: Fruit and seeds will still be produced if a plant is not given _____ or a ......
i.e. Phosphorous or a phosphatic fertilizer ONLY helps, or encourages, ...
∴ It is NOT TRUE to say that fruit and seeds will not be produced if a plant is not given a phosphatic fertilizer.

(h) Making plant tissues stronger helps a plant ...... (23)
If a plant is able to withstand mechanical damage, the entry of _____ such as bacteria and fungi is _____. (24)
i.e. Help in withstanding ...... prevents the _____ of pathogens into _____.
Help in withstanding mechanical damage prevents the entry of pathogens into plants.

EXERCISE A *Contextual reference*

Rewrite the following sentences replacing the words and phrases with the idea or expression they refer to in the text.

EXAMPLE

Sentence 4, *it*:

Farmyard manure contains three most important substances for plant materials – nitrogen, phosphate and potash.

1. Sentence 6, *it*
2. Sentence 8, *it*
3. Sentence 10, *this*
4. Sentence 11, *such crops*
5. Sentence 14, *they*
6. Sentence 16, *this element*
7. Sentence 20, *this element*
8. Sentence 23, *this*
9. Sentence 25, *so*
10. Sentence 25, *those*

EXERCISE B *Rephrasing*

Rewrite the following using other words and constructions from the text where possible.

1. A mixture of litter, solid excreta and urine improves the physical condition of the soil and makes up for lost plant nutrients.
2. The most commonly used fertilizer which contains nitrogen is made by combining ammonia with sulphuric acid.
3. When applied to the soil, the product which results from treating mineral phosphate with sulphuric acid promotes the production of fruit and seeds.
4. Broken branches and torn leaves allow pathogens such as bacteria and fungi to enter the plant and destroy it.
5. Essential elements of plant food are not readily available in a soil with a pH value of less than 5·0.

EXERCISE C *Relationships between statements*

In this book so far you have practised using a number of expressions that relate sentences together. These words indicate certain relationships between statements. These relationships are summarized in the table below. Copy out the table and complete it from the list of connectives provided.

Relationship	*Connectives*	
	Within sentences	*Between sentences*
CONSEQUENCE (Unit 1)	A and so B	so consequently ....
TIME SEQUENCE (Unit 2)	when after	firstly then ....
CONTRAST (Unit 2)	although	but however ....
ADDITION: REINFORCEMENT (Unit 3)	and	and moreover ....
ADDITION: SIMILARITY (Unit 3)		similarly ......
EXEMPLIFICATION (Unit 4)	for example	for example ......
EXPLANATION (Unit 4)	, or that is	that is to say ....

in other words	furthermore	therefore	also
in the same way	in contrast	such as	before
by this we mean	subsequently	B since A	hence
on the other hand	in addition	whereas	thus
as soon as	for instance	finally	

Choose the most appropriate expression from those given to relate the following sentences. Write out all the sentences. Replace and re-order the words in the sentences where necessary.

(a) 5, 6+7: therefore, furthermore, however
(b) 11+12: similarly, therefore, however, for example
(c) 13+14: thus, moreover, in contrast, also
(d) 22, 23+24: in other words, similarly, consequently, moreover
(e) 25+26: by this we mean, for example, but, also
(f) 30: although, that is, whereas, and so
(g) 30+31: also, hence, but, for instance
(h) 30, 31+32: that is to say, thus, however, so
(i) 34+35: thus, however, whereas, furthermore

II LANGUAGE IN USE

EXERCISE A *Conclusions based on observations*

Study the following information:

Mineral deficiencies

Soils are sometimes short of plant nutrients:

Symptom
plant stunted in growth, leaves yellowish in colour

Diagnosis
soil deficient in nitrogen

We can express this information as follows:

OBSERVATION		CONCLUSION
The plant is stunted in growth and the leaves are yellowish in colour.	This shows that	the soil is deficient in nitrogen.

= The plant is stunted in growth and the leaves are yellowish in colour. This shows that the soil is deficient in nitrogen.

Refer to your textbooks if necessary and show which symptoms go with which diagnoses in the following table by writing conclusions based on observations as in the example above. Add the following kinds of words where appropriate:

(a) definite and indefinite articles;
(b) forms of the verb *be*, and *there is/are*;
(c) *and* and *but*.

Symptom	*Diagnosis*
(a) plant roots dying at tips, or remain short and stubby	sulphur deficiency in soil
(b) all parts of plant stunted in growth, with bluish-green leaves, poor fruit or seed development	plant short of element magnesium
(c) plant tissues weak, prone to attack by insects and fungus	soil deficient in potassium
(d) leaves losing their green colour, becoming yellow at tips, between veins	deficiency of phosphorous in soil
(e) new leaves turning yellow, roots and stems becoming long and woody	supply of calcium in soil is low
(f) young leaves developing chlorosis, veins remain green	amount of zinc in soil insufficient
(g) foliage of field crops, particularly lower leaves, has intraveinal chlorosis	plant suffering from deficiency of iron

NOTE There are other ways of expressing the relationship between observations and conclusions:

Observation		*Conclusion*
The plant is stunted in growth and the leaves are pale yellowish in colour.	This suggests that ...... This indicates that ...... This is an indication that ...... This is a sign that ...... We may conclude that ...... This means that ...... So ...... Therefore	the soil is deficient in nitrogen.

EXERCISE B *Generalizations*

If we draw a conclusion from a number of observations we make a *generalization.*

If a plant is stunted in growth and the leaves are yellowish in colour, *then* the soil is deficient in nitrogen.

OR If plants are stunted in growth and the leaves are yellowish in colour, then the soil is deficient in nitrogen.

When a plant is stunted in growth and the leaves are yellowish in colour, the soil is deficient in nitrogen.

OR When plants are stunted in growth and the leaves are yellowish in colour, the soil is deficient in nitrogen.

Change the observations and conclusions you have written in Exercise A into generalizations.

EXERCISE C *Recommendations*

Observations + conclusions and generalizations can be followed by *recommendations* which indicate what remedy can be taken. For example:

Symptom	*Diagnosis*	*Remedy*
plant stunted in growth, leaves yellowish in colour	soil deficient in nitrogen	add nitrogenous fertilizer such as urea
OBSERVATION	CONCLUSION	RECOMMENDATION
The plant is stunted in growth and the leaves are yellowish in colour.	The soil is deficient in nitrogen.	Add nitrogenous fertilizer such as urea.

(i) *observation + conclusion + recommendation*
The plant is stunted in growth and the leaves are yellowish in colour. This shows that the soil is deficient in nitrogen. *This deficiency can be corrected by adding a nitrogenous fertilizer such as urea.*

OR ...... *In order to correct this deficiency, a nitrogenous fertilizer such as urea can be added to the soil.*

(ii) *generalization + recommendation*
If plants are stunted in growth and the leaves are yellowish in colour, then the soil is deficient in nitrogen. *This deficiency can be corrected by adding a nitrogenous fertilizer such as urea.*

OR ...... *In order to correct this deficiency* ...... etc.

NOTE Instead of *correct/corrected* we can use *remedy/remedied.*

Part 1

Add appropriate recommendations, as in the examples above, to (i) the observations + conclusions you have made in Exercise A, and (ii) the generalizations you have made in Exercise B.

(a) spread lime over the land
(b) apply a phosphatic fertilizer such as superphosphate
(c) add a potassic fertilizer such as muriate of potash
(d) add magnesium lime or sulphate of potash to the soil
(e) add sulphate of ammonia to the soil
(f) add ferric oxide to the soil
(g) spray zinc sulphate on the soil, or mix it with the fertilizer

Part 2

A recommendation can be made stronger by using *should* instead of *can.*

EXAMPLE

> The plant is stunted in growth and the leaves are yellowish in colour. This shows that the soil is deficient in nitrogen. This deficiency *should* be corrected by adding a nitrogenous fertilizer such as urea.

Make some of the recommendations you have made in Part 1 above stronger by using *should* instead of *can.*

EXERCISE D *Predictions*

The observations practised in Exercise A can also be used as *predictions* following a statement expressing a condition. A prediction is a statement about what will happen (i.e. in the future) if a situation occurs. Study the following information and the way it can be expressed.

Mineral deficiency	*Symptom*
nitrogen	plants stunted in growth, leaves yellowish in colour
CONDITION	PREDICTION
A soil is deficient nitrogen.	Plants will be stunted in growth and the leaves will be yellowish in colour.

= *If* a soil is deficient in nitrogen, plants *will be* stunted in growth and the leaves yellowish in colour.

Such *conditions+predictions* can be followed by recommendations as in Exercise C above.

EXAMPLE

If a soil is deficient in nitrogen, plants will be stunted in growth and the leaves yellowish in colour. *This deficiency can/should be corrected/remedied by adding a nitrogenous fertilizer such as urea.*

OR If a soil is deficient in nitrogen, plants will be stunted in growth and the leaves yellowish in colour. *In order to correct/remedy this deficiency, a nitrogenous fertilizer such as urea can/should be added to the soil.*

Part 1

Express the following information as conditions + predictions followed by recommendations as in the example above. Add the following kinds of words where appropriate:

(a) definite and indefinite articles;
(b) forms of the verb *be*;
(c) *and* and *but*.

CONDITION + PREDICTION		RECOMMENDATION
Mineral deficiency	*Symptom of deficiency*	*Remedy*
1. calcium	plant roots die at tips, or remain short and stubby	spread lime over land
2. phosphorous	all parts of plant stunted in growth, with bluish green leaves, poor fruit or seed development	apply phosphatic fertilizer e.g. superphosphate
3. potassium	plant tissues weak, prone to attack by insects and fungus	add potassic fertilizer like muriate of potash
4. magnesium	leaves lose their green colour, become yellow at tips, between veins	add magnesium lime or sulphate of potash to soil
5. sulphur	new leaves turn yellow, roots and stems become long and woody	add sulphate of ammonia to soil
6. iron	young leaves develop chlorosis, veins remain green	add ferric oxide to soil
7. zinc	foliage of field crops, particularly lower leaves, has intraveinal chlorosis	spray zinc sulphate on soil or mix it with fertilizer

Part 2

There are two other ways of expressing a condition:

EXAMPLES

If there is a deficiency of nitrogen in the soil, plants will be ......
If the supply of nitrogen in the soil is low, plants will be ......

Rewrite some of the conditions + predictions followed by recommendations you have made in Part 1 expressing the condition in the ways shown in the examples above.

III GRAMMAR

EXERCISE A *Defining and non-defining relative clauses*

1. Look at these diagrams:

A		+	B		=	C	
Fertilizers			Fertilizers			Fertilizers	
/	\		/	\		/	\
Nitrogenous	Phosphatic		Have nitrogen as main element	Have phosphorous as main element		Nitrogenous: have nitrogen as main element	Phosphatic: have phosphorous as main element

[A] Some fertilizers are called nitrogenous fertilizers.

[B] Some fertilizers have nitrogen as the main food element.

[A+B] = [C] Fertilizers *which have nitrogen as the main food element* are called nitrogenous fertilizers.

OR Nitrogenous fertilizers are fertilizers which have nitrogen as the main food element.

[A] Some fertilizers are called phosphatic fertilizers.

[B] Some fertilizers have phosphorous as the main food element.

[A+B] = [C] Fertilizers *which have phosphorous as the main food element* are called phosphatic fertilizers.

OR Phosphatic fertilizers are fertilizers which have phosphorous as the main food element.

Here the clauses *in italics* define the kind of fertilizer: they are *defining relative* clauses.

Now look at these diagrams:

Fertilizers

A	Nitrogenous	Phosphatic
B	Have nitrogen as main element	Have phosphorous as main element
C	Strengthen plant tissues	Improve plant growth

A(+B)+C Nitrogenous fertilizers, *which have nitrogen as the main food element*, strengthen plant tissues.
Phosphatic fertilizers, *which have phosphorous as the main food element*, improve plant growth.

Here the clauses *in italics* give us additional information about the fertilizers. They are *non-defining relative* clauses. They are separated from the rest of the sentences by commas.

Combine each pair of sentences into a single sentence. Change the second sentence into a relative clause and insert it into the first sentence at the point indicated by the dots. In each case say whether the relative clause is defining or non-defining.

1. A leguminous crop ...... will add as much nitrogen to the soil per acre as 3 to 10 tons of farmyard manure. A leguminous crop is ploughed under.
2. Solid excreta, or faeces, is the material ...... The material has passed through the animal without being digested.
3. Ammonium sulphate ...... is the most commonly used nitrogenous fertilizer. Ammonium sulphate supplies the soil with nitrogen and sulphur.
4. When crude salt is purified it is called muriate of potash ...... Muriate of potash contains 50–60% K_2O.
5. Another fertilizer ...... is urea. This fertilizer contains nitrogen.
6. A soil ...... will contain quite a high proportion of calcium. Such a soil has a pH value of 6·5 or more.
7. The roots of leguminous crops such as clover bear nodules ...... The nodules contain bacteria ...... These bacteria accumulate nitrogen from the air.
8. A compost is a mixture of partly broken down material ...... This material is usually made up of leaves or grass cuttings.

9. Fungi ...... can be controlled by means of chemical substances ...... Fungi attack the aerial parts of the crop, the leaves, stems etc. These chemical substances are known as fungicides.
10. In addition to the method of disease control ...... there are other methods of control ...... Disease control attacks the disease organism. These other methods arc particularly important in crop production.
11. The quantity of fertilizer or manure ...... depends on the fertility of the soil and the nutrients ...... Fertilizer or manure is required for rice cultivation. The nutrients are released from the rotted weeds and stubble ...... The weeds and stubble are ploughed in during cultivation.
12. The type and amount of fertilizer required also depends on the variety of rice used, and certain varieties ......, ..., will often respond better to high levels of fertilizer than late maturing varieties. Certain varieties have short erect leaves. These varieties include many of the early maturing varieties. (omit *certain*)

EXERCISE B *Short-form relative clauses*

Relative clauses often appear in a shortened form.

(i) In clauses which begin *which has* or *which have*, *with* may be used instead:

EXAMPLE

Rice varieties *which have short erect leaves* respond well to high levels of fertilizer.

= Rice varieties *with short erect leaves* respond well to high levels of fertilizer.

(ii) In clauses in which the verb is active the *-ing* form of the verb may be used instead.

EXAMPLES

Rice varieties *which have short erect leaves* respond well to high levels of fertilizer.

= Rice varieties *having short erect leaves* respond well to high levels of fertilizer.

The most commonly used fertilizer *which contains* nitrogen is ammonium sulphate.

= The most commonly used fertilizer *containing* nitrogen is ammonium sulphate.

(iii) In clauses in which the verb is passive, the relative pronoun and the form of verb *to be* can be omitted.

EXAMPLES

The quantity of fertilizer or manure *which is required for rice* cultivation partly depends on the variety of rice *which is used.*

= The quantity of fertilizer *required for rice cultivation* partly depends on the variety of rice *used.*

(iv) In clauses in which the relative pronoun is followed by a form of the verb *to be* and an adjective or a noun the relative pronoun and the form of verb *to be* can be omitted.

EXAMPLES

Common nitrogen fertilizers *which are suitable for rice* are ammonium sulphate and urea.

= Common nitrogen fertilizers *suitable for rice* are ammonium sulphate and urea.

Urea, *which is the substance in human and animal urine,* is a nitrogenous fertilizer.

= Urea, *the substance in human and animal urine,* is a nitrogenous fertilizer.

Part 1

Combine each of the following pairs of sentences into a single sentence. Change the second sentence into a short-form relative clause, using one of the ways shown above, and insert it into the first sentence at the point indicated by the dots. In each case state whether the resulting construction is defining or non-defining, as in the example.

EXAMPLE

Muriate of potash is a form of potash ...... This form of potash is used in many compound fertilizers.

= Muriate of potash is the form of potash *used in many compound fertilizers.* (defining)

1. Viruses are very small organisms ......
   They are usually transmitted by means of insects.
2. Nematodes are small worms in the soil that enter plant roots ......
   They cause serious losses in some crops, particularly in the tropics.
3. Compound fertilizers are multiple nutrient materials ......
   They supply two or three plant nutrients simultaneously.
4. It is essential to understand the materials ......
   The materials are available to the farmer to maintain the supply of minerals in the soil.

5. Some fertilizers ...... should be applied by means of a drill in 1–2″ wide bands on either side of a row of seeds at a depth of 4–6″.
   Some fertilizers contain soluble phosphates. (omit *some*)
6. Fungi are microscopic organisms ......
   These organisms produce thread-like growths ......
   These growths arc known as hyphae.
7. The types of disease organisms ...... are fungi and bacteria.
   They are found in the soil ......
   They attack the roots of plants and they cause root rot and wilt.
8. Christmas Island Rock Phosphate, ..., also contains calcium, ...
   This type of phosphate is the most popular phosphorous-containing fertilizer.
   Calcium is essential for growing crops in the wet tropics.
9. Mixed fertilizers ...... usually meet nutrient deficiencies in a more balanced manner and require less labour to apply than straight fertilizers ......
   Mixed fertilizers contain all the three principal nutrients (N, P and K).
   Straight fertilizers are used separately.
10. A soil ...... would contain quite a high proportion of calcium.
    However, a soil ...... would be considered strongly acid.
    High acidity means low calcium, magnesium and potassium, ...
    A soil has a pH value of 6·5 or more.
    A soil has a pH value ......
    The value is lower than 5.
    Calcium, magnesium and potassium are all important plant foods.

Part 2

Look at the sentences containing relative clauses which you wrote in Exercise A. Shorten the following sentences from this exercise in one of the ways shown on pages 70 and 71:

5, 6, 8, 9, 10, 11, 12.

IV SUMMARY CHECK

Complete the following text by filling in the blank spaces. Some of the expressions you will require are given below. A dotted line ...... requires a phrase to be added and a straight line _____ requires a word.

therefore	constituent	deficient in
essential elements	plant food	plant nutrition
soil fertility	calcium pectate	stunted growth
nitrogen	promotes	nutrients
ammonium sulphate	potassium	acid

Fertilizers are of crucial importance in maintaining and improving ...... by ensuring an adequate supply of plant _____ for satisfactory plant _____. For example, nitrogen promotes rapid _____ and gives plants a healthy green colour. If a soil is deficient in _____ plants will be stunted in _____ and the leaves yellowish in colour. However, this _____ can be corrected by adding to the soil a ...... such as urea. To take another example: phosphorous stimulates early growth and root formation, and _____ fruit and seed production. If there is a _____ of _____ in the soil, plants will be stunted in growth with bluish green leaves and poor ...... development. To remedy this situation a ...... such as superphosphate should be applied.

The most commonly used nitrogenous fertilizer is ammonium sulphate which supplies the soil with both _____ and _____. There are other fertilizers which contain nitrogen including urea, which is made from ammonia and carbon dioxide. _____ contains about 46% _____, more than double the amount in ...... It is, _____, a valuable fertilizer to use in soils which are ...... nitrogen, as in the tropics, for example.

One result of nutrient deficiency is an increase in soil acidity. The effect of _____ is to make certain elements unavailable to plants. To correct this, lime should be _____. The main _____ of lime is calcium, itself an essential ...... which combines with pectin in plants to form ......, an essential element of cell walls. But the main effect of lime is to make ...... of plant food available to plants. Thus, _____, phosphorous and _____ are more easily available in a well-limed soil than in an _____ soil.

6 The Control of Weeds and Plant Diseases

I READING AND COMPREHENSION

[1]In crop production the control of weeds, diseases and pests is essential to obtain high yields. [2]All three may be controlled by sound farm practices. [3]These include the choice of clean seed and the growing of varieties of crop which can resist disease. [4]They also include careful cultivation, both pre-sowing and post-sowing, and the use of chemicals.

[5]Weeds reduce crop yields on account of the fact that they compete with crops for water, soil nutrients and light. [6]They also make harvesting difficult. [7]Most weeds are aggressive and invasive, they grow quickly and spread far, and so are difficult to get rid of. [8]One recommended way of eradicating many persistent weeds is first to plough up the roots and underground parts of the plant. [9]Then the soil may be cultivated lightly, or rotavated, on one or more occasions after the first ploughing.

(a) One sound farm practice is the choice of disease resistant crop varieties.
(b) Aggressive and invasive weeds cannot be eradicated.

[10]The principal reason for cultivating the soil is to kill weeds. [11]Weeds may also be killed by means of chemicals which have the collective name of herbicides. [12]Weed-killers are of two basic types: selective and non-selective. [13]The former remove certain weeds from certain crops. [14]For rice we can spray the herbicide 2:4-D or MCPA over the whole crop at low concentrations ($\frac{1}{2}$–1 lb. per acre). [15]The rice will not be affected, but many of the rice weeds will be killed. [16]Non-selective weed killers may be used for removing all vegetation e.g. as brush killers. [17]They must be used extremely carefully for the simple reason that they will eradicate all plants on contact – which includes the crop itself. [18]They are usually used before sowing or before the emergence of the crop itself.

(c) Selective weed killers do not affect the crop which is being grown.
(d) Non-selective weed-killers eradicate certain kinds of weeds.

[19]Plant diseases are caused by organisms which use the crop plant as a 'host'. [20]These are mainly micro-organisms e.g. fungi, bacteria and viruses.

[21]These parasitic micro-organisms live off the food nutrients in the tissue cells of the plants. [22]They frequently kill the host tissues, and either the whole plant or a part of it is damaged and killed. [23]Micro-organisms are reproduced and spread by minute bodies such as spores, fungi and bacteria. [24]Wind, water, diseased plants, cuttings and tubers, animals, men and insects are some of the means whereby disease is disseminated.

[25]It is very difficult to kill the fungi and bacteria, or to make the virus which is inside the host plant inactive. [26]But the evolution of plant varieties which can resist disease has completely changed methods of disease control. [27]A number of varieties have been evolved and are now available to farmers. [28]So the control of plant diseases has increasingly become a matter of prevention.

(e) By using the plant as 'host' bacteria cause plant diseases.
(f) Plants are damaged and killed because micro-organisms live off the tissue cells of the plants.
(g) A number of disease resistant plant varieties have been evolved.

[29]Fungi, which attack the aerial parts of the crop, can be controlled by means of fungicides. [30]These are sprayed or dusted on to the plant surfaces. [31]They should be applied before the plant is seriously damaged. [32]Sometimes spray and dust is applied whether disease is present or not. [33]In any case, it is necessary to examine crops frequently for signs of disease.

[34]Soil-borne diseases are much more difficult to control. [35]There are various ways of treating the soil. [36]One way is to use chemicals that easily change into a gas or vapour, which enter the soil and kill the harmful organisms. [37]The soil is covered with a polythene sheet and the volatile chemical is injected into the soil. [38]After about 24 hours the sheet is removed and the soil is allowed to air for a few days before use.

Solutions

(a) Weeds, diseases and pests may be controlled by . . . . . . (2)
Sound farm practices *include* the choice of clean seed and . . . . . . (3)
i.e. The growing of varieties of crops which can resist disease is *one example* of a . . . . . .
= One sound farm practice is the _____ of disease resistant . . . . . .
One sound farm practice is the choice of disease resistant crop varieties.

(b) Most weeds grow quickly and spread far.
i.e. They are _____ and _____. (7)
Aggressive and invasive weeds are difficult . . . . . .
but: There are certain recommended ways of getting rid of _____ weeds. (8)
i.e. Persistent weeds (i.e. . . . . . . weeds) CAN be got rid of (i.e. _____). (8)
Aggressive and invasive weeds CAN be eradicated.

(c) We can spray rice with ...... (14)
2:4-D or MCPA are ____ weed-killers. (12)
Selective weed-killers remove ...... (13)
i.e. Many of the rice weeds will be killed, but the rice ...... (15)
i.e. ......do not affect the crop which is being grown.
Selective weed-killers do not affect the crop which is being grown.

(d) Selective weed-killers remove ...... (13)
but: ALL weeds will be removed from ALL crops by ...... (16)
i.e. Non-selective weed-killers remove, or get rid of, or ____, ALL kinds of weeds.
Non-selective weed-killers eradicate ALL kinds of weeds.

(e) Micro-organisms which use the crop plant as a 'host' cause ...... (19)
i.e. Plant diseases are caused by micro-organisms such as ...... (20)
∴ Fungi, bacteria and viruses cause plant diseases by using ......
= By using the crop plant as a 'host' fungi, ...
∴ *By using the crop plant as a 'host' bacteria cause plant diseases.*

(f) The tissue cells of the plant provide food nutrients for micro-organisms to ...... (21)
As a result of micro-organisms living off ...... of the plant, either the whole plant or a part of it is ......
i.e. Plants are damaged and killed because ...... the tissue cells of the plants.
Plants are damaged and killed because micro-organisms live off the tissue cells of the plants.

(g) Plant varieties have been evolved which can ...... (26)
Plant varieties which can resist disease = plant varieties which are ...... (see solution **a**)
i.e. A number of disease resistant ...... have been ____.
A number of disease resistant plant varieties have been evolved.

EXERCISE A *Contextual reference*

Rewrite the following sentences replacing the words and phrases with the idea or expression they refer to in the text.

1. Sentence 2, *all three*
2. Sentence 3, *these*
3. Sentence 4, *they*
4. Sentence 5, *they*
5. Sentence 6, *they*
6. Sentence 13, *the former*
7. Sentence 17, *they* ...... *they*

8. Sentence 20, *these*
9. Sentence 22, *they*
10. Sentence 27, *varieties*
11. Sentence 31, *they*

EXERCISE B *Rephrasing*

Rewrite the following using other words and constructions from the text where possible.

1. Weeds which grow quickly and spread far are difficult to get rid of.
2. A way which is frequently advised to eradicate weeds which last a long time is to plough first and then cultivate lightly.
3. Chemicals which remove certain weeds from certain crops are used in rice cultivation, when they are sprayed over the whole crop at low concentrations.
4. Organisms which are microscopic in size and use the crop plant as a 'host' are frequently the cause of a whole crop being killed.
5. Plant diseases are spread in a variety of ways – by wind, water, animals, men and insects, for instance.
6. In order to control soil-borne diseases a chemical that easily changes into a gas or vapour can be injected into the soil under a polythene sheet.

EXERCISE C *Relationships between statements*

Choose the most appropriate expressions from those given to relate the following sentences. Write out all the sentences. Replace and re-order the words in the sentences where necessary.

(a) 2+3+4: such as, however, therefore, moreover
(b) 5: because, whereas, as soon as, such as
(c) 7: on the other hand, because, in addition, that is to say
(d) 8+9: however, consequently, subsequently, moreover
(e) 10+11: consequently, moreover, whereas, however
(f) 12, 13+14: therefore, for example, moreover, subsequently
(g) 13, 14, 15+16: therefore, moreover, in contrast, as a result
(h) 16+17: therefore, whereas, for example, moreover
(i) 17+18: but, also, thus, and
(j) 21+22: moreover, however, as a result, in addition
(k) 25+26: consequently, moreover, therefore, however
(l) 27+28: for example, thus, subsequently, finally.

II LANGUAGE IN USE

EXERCISE A *The identification and description of diseases*

Part 1

Crop diseases may be identified by name and by the organism that causes the disease.

EXAMPLE

Crop	*Name of disease*	*Causal organism*
Rice	blast	fungus: *Piricularia oryzae*

We can make statements of identification as follows. Note that the form of some of the statements is the same as that used to express definitions. (See Unit 1, Language in Use, Exercise A, and Unit 1, Grammar)

Blast in rice is a fungus disease (which is) caused by the organism *Piricularia oryzae*.

OR The fungal organism *Piricularia oryzae* causes/is the cause of blast in rice.

OR *Piricularia oryzae* is a fungal organism which causes blast in rice.

OR The cause of blast is rice is the fungus/fungal organism *Piricularia oryzae*.

Make statements of identification about the following plant diseases in the ways shown above.

Crop	*Name of disease*	*Causal organism*
1. Maize	smut	fungus: *Ustilago zeae*
2. Cotton	stenosis, or leaf curl	virus
3. Cotton	black arm, or angular leaf spot	bacteria: *Xanthomas malvacearum*
4. Sugar	red rot	fungus: *Colletotrichum falcatum*

Crop	*Name of disease*	*Causal organism*
5. Sweet potatoes	black rot	fungus: *Ceratostomella funbriata*
6. Groundnuts	rosette disease	virus
7. Tomatoes	bacterial wilt	bacteria: *Pseudomonas solanaceum*

Part 2

We can make statements *describing* the symptoms of particular diseases.

EXAMPLE

Crop	*Name of disease*	*Causal organism*	*Symptoms*
Rice	blast	fungus: *Piricularia oryzae*	Brown longitudinal spots on leaves. Spots on stem and grain darker in colour

The statements describing the disease follow the statement identifying it.

EXAMPLES

Blast in rice is a fungus disease caused by the organism *Piricularia oryzae*. Brown longitudinal spots *appear* on the leaves. The spots on the stem and grain are darker in colour.

OR The fungal organism *Piricularia oryzae* is the cause of blast in rice. Brown longitudinal spots *can be seen* on the leaves. The spots on the stem and grain are darker in colour.

Using the information in the table on the next page, add statements describing the symptoms of the disease to the statement of identification you made in Part 1. Add the following kinds of words, where appropriate:

(a) definite and indefinite articles;
(b) forms of the verb *be* and *appear*, *can be seen*;
(c) *and*.

IDENTIFICATION			DESCRIPTION
Crop	*Name of disease*	*Causal organism*	*Symptoms*
1. Maize	smut	fungus: *Ustilago zeae*	Soft tumours on all parts of plant, mostly cob. Tumours, when mature, split and release black dusty mass of spores.
2. Cotton	stenosis, or leaf curl	virus	Leaves of affected plant become crinkled and deformed. Yellow or pink discoloration.
3. Cotton	black arm, or angular leaf spot	bacteria: *Xanthomonas malvacearum*	All above ground parts of plant attacked. Angular spots on leaves and branches turn dark brown with reddish margin when old.
4. Sugar cane	red rot	fungus: *Colletotrichum falcatum*	Leaves wither, cane shrinks. Black specks on shrivelled rind. Split cane gives sour smell.
5. Sweet potatoes	black rot	fungus: *Ceratostomella funbriata*	Tubers attacked. Foliage turns yellow and sickly. Black cankers on portion underground. Grey-black, circular, depressed spots on fleshy roots.
6. Groundnuts	rosette disease	virus	Circular spots on both sides of leaves. Spots on upper surface dark brown with yellow halo.
7. Tomatoes	bacterial wilt	bacteria: *Pseudomonas solanacerum*	Lower leaves wilt, and eventually die.

EXERCISE B *Recommendations*

After identifying the disease and describing it, we can make statements *recommending* particular control measures or ways of avoiding the disease.

EXAMPLE

Crop	*Name of disease*	*Causal organism*	*Symptom*	*Control measures*
Rice	blast	fungus: *Piricularia oryzae*	Brown longitudinal spots on leaves. Spots on stem and grain darker in colour.	Spray with 1% Bordeaux mixture. Also/Or grow resistant varieties.

We can express recommendations in various ways:

As a control measure, the crop should be sprayed with 1% Bordeaux mixture. *In addition,/Alternatively, to avoid the disease,* resistant crop varieties should be grown.

OR *One control measure is to* spray with 1% Bordeaux mixture. *In addition/ Alternatively,* resistant crop varieties should be grown.

OR *Control measures for this disease include* spraying with 1% Bordeaux mixture, and growing resistant crop varieties.

Using the information in the table below, add statements recommending control measures or ways of avoiding the disease, as in the example above, to the statements of identification and description you made in Part 2 of Exercise A. In addition to the expressions printed in italics in the examples above, add definite and indefinite articles.

EXAMPLES

Blast in rice is a fungus disease caused by the organism *Piricularia oryzae*. In this disease brown longitudinal spots appear on the leaves. The spots on the stem and grain are darker in colour. As a control measure the crop can be sprayed with 1% Bordeaux mixture. To avoid the disease, resistant crop varieties should be grown.

OR The fungal organism *Piricularia oryzae* causes blast in rice. In this disease brown longitudinal spots can be seen on the leaves. The spots on the stem and grain are darker in colour. One control measure is to spray the crop with 1% Bordeaux mixture. Alternatively, resistant crop varieties can be grown.

OR The fungal organism *Piricularia oryzae* is the cause of blast in rice. In this disease brown longitudinal spots can be seen on the leaves. The spots on the stem and grain are darker in colour. Control measures for this disease include spraying with 1% Bordeaux mixture and growing resistant crop varieties.

Study the following table and write different paragraphs, as in the examples above, *identifying* and *describing* the disease and *recommending* control measures or ways to avoid it. Add any other words and phrases which are necessary, e.g. definite and indefinite articles, the verb *to be* etc.

IDENTIFICATION			DESCRIPTION	RECOMMENDATION
Crop	*Name of disease*	*Causal organism*	*Symptoms*	*Control measures*
1. Maize	smut	fungus: *Ustilago zeae*	Soft tumours on all parts of plant, mostly cob. Tumours when mature, split and release black dusty mass of spores.	Practise crop rotation and sanitation. Or, grow resistant varieties.
2. Cotton	black arm, or angular leaf spot	bacteria: *Xanthomonas malvacearum*	All above ground parts of plant attacked. Angular spots on leaves and branches turn dark brown with reddish margin when old.	Treat seed with organo-mecurials (Agroson 5 W) before sowing. Also, destroy all cotton plant debris.
3. Sugar cane	red rot	fungus: *Colletotrichum falcatum*	Leaves wither, cane shrinks. Black specks on shrivelled rind. Split cane gives sour smell.	Plant clean cuttings dipped in 5% formalin. Or, grow resistant varieties.
4. Sweet potatoes	black rot	fungus: *Ceratostomella funbriata*	Tubers attacked, foliage turns yellow and sickly. Black cankers on portion underground. Grey-black, circular depressed spots on fleshy roots.	Treat seed tubers in mercuric chloride solution, or 2½ borax solution. Also, carefully store seed tubers, and rotate seed-bed.
5. Groundnuts	rosette disease	virus	Circular spots on both sides of leaves. Spots on upper surface dark brown with yellow halo.	Plant seeds early and close. Also, use clean seed, and uproot and burn infected plants.

III GRAMMAR

EXERCISE A *Noun+noun constructions*

An adjective is often used to modify a noun. For example:

nitrogenous fertilizer	*volatile* chemical
selective weed-killer	*leguminous* crop
parasitic micro-organisms	*organic* material

A noun may also be used to modify another noun. Many different kinds of relationship are possible in such *noun+noun* constructions. Study the following carefully:

A made from B:	silage made from an arable crop = *arable silage*
A caused by B:	a disease which is caused by a fungus = *fungus disease*
A used for B:	tiles used for drainage = *drainage tiles*
	fans used for winnowing = *winnowing fans*
A in/on B:	poisons which act in the stomach of a pest = *stomach poisons*
A part of B:	the sap of the cell = *cell sap*
A shaped like B:	a root shaped like a club = *club root*
A operated by B:	a mill which is operated by wind = *windmill*
A which does B:	a substance which kills weeds = *weed-killer*
	a machine which seeds rice = *rice seeder*

Notice that we can often write a simple definition of each noun+noun compound by means of a statement which makes the relationship between the nouns clear.

EXAMPLES

Arable silage is silage (which is) made from an arable crop.
A fungus disease is a disease (which is) caused by a fungus.
Drainage tiles are tiles which are used for drainage of land.
Stomach poisons are poisons which act in the stomach of a pest.

Now write statements which show the full meaning of the following *noun+noun* compounds. If possible write definitions.

1. farm implements
2. plantation crops
3. asbestos sheet
4. leaf cell
5. root hairs
6. virus disease
7. dusting equipment
8. soil injector
9. farmyard manure
10. cattle dung
11. fodder crops
12. maize sheller
13. soil humus
14. grinding mill
15. rice huller
16. herring-bone system
17. lime nodules
18. threshing machine

19. grain thresher
20. seed drill
21. poultry manure
22. sugar cane crusher
23. earthworms
24. spring wheat
25. grazing land
26. forage crops
27. wire netting
28. buffalo plough
29. soil water
30. leaf blade

EXERCISE B *Particle + noun constructions*

Apart from *adjective + noun* and *noun + noun* phrases there are a large number of modifiers which are made up of a *verb stem + suffix*. Two important classes of modifier are as follows:

(i) Modifiers formed from verbs by adding *-ing* (present participles). These modifiers are usually active in meaning.

EXAMPLES

processing machines = machines which process something
threshing implements = implements which thresh some crop

In addition, the object of the verb may also be included:

soil transporting implements = implements which transport soil

(ii) Modifiers formed from verbs by using the past participle form. (e.g. transported, made, known etc.) These modifiers are usually passive in meaning.

EXAMPLES

ground chalk = chalk which has been ground (to a powder)
transported soil = soil which {has been / is being} transported

In addition, the agent which performs the action described by the verb may be included, often connected by a hyphen:

air-borne diseases = diseases which are borne by air

Now write down and complete the sentences by filling in the blanks with a past participle modifier or a present participle modifier. Form each modifier from one of the verbs in this list.

fix	spray	lime	operate	recommend
drive	cultivate	dust	bear	
draw	lift	conduct	compose	

1. Leguminous plants have nitrogen ...... bacteria growing on their roots, which extract nitrogen from the soil air, use what they need themselves, and pass on the rest to the host plant.
2. For carrying out heavy jobs which need a lot of power, such as ploughing or land reclamation, tractor ...... implements are superior in every way to manually ...... or bullock ...... implements.

3. Nitrogen, phosphorous and potassium are more easily available to plants in a well-...... soil than in a soil which has a low pH value and is therefore acid.
4. If the mid-rib of a leaf is examined under a microscope it will be seen to consist of water and food ...... tissues called *xylem* and *phloem* respectively.
5. Due to the fact that grasslands have diminished in many countries in recent years ...... fodder crops have become increasingly important, particularly for dairy cattle.
6. The Persian Wheel, or *rahet*, one of the most common of water-...... appliances, consists of a series of baskets on an endless belt which moves on a vertical wheel or drum.
7. In the case of soil-...... diseases where the fungus living in the soil attacks the seeds or seedlings, soil disinfectants or sterilizers such as formaldehyde or formasan are helpful in destroying the fungus.
8. One of the ...... methods of eradicating the paddy stem borer is dipping the seedlings in 0·1% DDT suspension before transplanting.
9. As most vegetable crops are usually affected by a variety of fungus and virus diseases and a number of insect pests, it is better for the farmer to equip himself with the necessary ...... and ...... machines.
10. ...... manure, or compost, can be made from different sorts of waste material including padi-straw, grass clippings, sugar-cane refuse, etc.

EXERCISE C *Complex noun phrases*

The names of agricultural tools and equipment are often long and complex noun phrases.

EXAMPLE

This plough is used for ridging. It has high wings which are adjustable. It can be drawn by an animal.

= an animal-drawn adjustable high-wing ridging plough.

Look at the following sentences describing a piece of agricultural equipment and decide which noun phrase is appropriate for naming it.

1. This harrow is toothed with sharp spikes. It is designed to be drawn by an animal. It is a spike-drawn, harrow-toothed animal/an animal-drawn spike-toothed harrow.
2. This hoe is made specially for African soils. The depth of work can be adjusted. It is an African adjusted hoe/an adjustable African hoe.
3. This drill is for seeding. It is designed to be pushed by hand. It is a seeding hand drill/a hand-pushed seed drill.
4. This seed drill is designed for animal draft. It works automatically. It is a drill-designed draft animal/an animal-drawn automatic seed drill.

5. This machine can be used to spray plantation crops by hand. The hand pump gives a continuous spray at high pressure. It is a continuous pressure hand spray/a high-pressure plantation crop-sprayer.
6. This sprayer is designed to be supported on the back like a knapsack. It sprays pneumatically. It is a pneumatic knapsack sprayer/a back-supported spraying knapsack.
7. This set is used for pumping water from deep wells. It can be operated by one man. It is a one-man deep-well operator/a manually-operated water pump.
8. This thresher, designed for threshing rice, can be operated by hand. It is a hand-operated rice thresher/a hand-designed rice thresher.

IV SUMMARY CHECK

Complete the following text by filling in the blank spaces. Some of the expressions you will require are given below. A dotted line ...... requires a phrase to be added and a straight line _____ requires a word.

seeds	borne	diseases	bacteria
chemical substances	inactivate	infection	viruses
available	disease-free	host plant	fungal organism
fungicide	microscopic	resistant	control measure
function	spraying or dusting	organisms	

Plant disease may be defined as a condition in which the plant as a whole, or any part of it, does not perform its normal _____. It may be brought about by a number of _____ which are all _____ in size such as fungi, _____, _____ etc., or by physiological causes. These _____ -organisms obtain their food in two ways: by breaking down dead plants or animal remains (saprophytes) or by attacking living plants and _____ (parasites).

Fungi are _____ producing thread-like growths known as *hyphae*. For example, the ...... *Piricularia oryzae* causes blast in rice. In this disease brown, longitudinal spots can be seen on the leaves. The spots on the stem and grain are darker in colour. Fungi can be _____ by means of ...... known as fungicides. There are many kinds of _____ but they are all used by ...... them on to the plant surfaces. One control measure against blast is to spray the crops with 1 % Bordeaux mixture. Alternatively, _____ crop varieties can be grown.

Bacteria are microscopic cellular _____. They occur, like fungi, both in the air and in the _____. For example, black arm, or angular leaf spot, is a bacterial _____ in cotton caused by the organism *Xanthomonas malvacearum*. All above ground parts of the plant are attacked. Angular spots appear on the leaves and the branches turn dark brown with a reddish margin when old. As a ...... the cotton seed should be treated with an organo-mecurial _____ (Agroson GN or Ceresan) before sowing. Such preparations have proved very effective for the control of seed-_____ and some soil-borne _____. They are, however, poisonous to men and beasts, and great care should be taken in handling these _____ and the _____ treated with them.

Virus_____ are usually spread by direct contact, through diseased cuttings and grafts and by insects. . . . . . . cannot be prevented or cured by chemicals. Thus, the use of _____ seed is essential to reduce _____. An example of a . . . . . . is Rosette in groundnuts. In this _____ circular spots occur on both sides of the leaves. The spots on the upper surface are dark brown with a yellow halo. To prevent the disease _____ should be planted early and close. Clean seed should be used and _____ plants uprooted and burnt.

It is very difficult to kill fungi and bacteria and to _____ virus once it is inside the . . . . . . Hence, the control of . . . . . . increasingly lies in preventing _____. The development of . . . . . . has, however, revolutionized methods of . . . . . . Nowadays, _____ of most of the economic crops which are _____ to certain important _____ are available to farmers.

7 Market Gardening

I READING AND COMPREHENSION

EXERCISE A *Re-ordering paragraphs*

Write out the following paragraphs in the order which corresponds to the following headings:

1. Advantages of market gardening
2. Importance of seed selection
3. Types of soil for market gardening
4. Preparation of the plot
5. Use of manures and fertilizers
6. Nursery beds
7. Transplantation

(a) The plot should be more or less level and laid out in beds. Narrow paths between the beds facilitate planting, watering, weeding and harvesting. Dry season beds slope inwards to hold water in the bed. Rainy season beds are ridged.‸

(b) The aim of transplanting is to give each plant more space to develop its roots and leaves. Planting distances vary from species to species.‸ The aim is to have a full crop cover of the ground when plants are mature.‸‸

(c) Market gardening is the cultivation of vegetables for sale at markets in towns. Vegetables are short-duration crops and all the family labour of the grower can be employed throughout the year.‸ Vegetables can usually be marketed at a good price. ‸

(d) Vegetables which have small seeds such as tomatoes, lettuce and cabbage are planted first in boxes or in special nursery beds to raise the seedlings. When they have grown to a suitable size they are transplanted to prepared beds. The soil should be clean and free from disease organisms. ‸There are various ways of getting rid of disease organisms.‸‸

(e) The land selected for market gardening should have a loose, friable, free-draining soil which does not easily get waterlogged. A loam or sandy loam is usually preferred.‸‸

(f) Vegetable crops are heavy feeders.‸ Depending on the soil a variable amount of organic manures should be applied. For a very sandy soil a higher proportion is used, about 1:3. Bag fertilizers should be used that contain all the major plant nutrients. For leafy vegetables a mixture with a high nitrogen content is best.ʌ

(g) Successful vegetable growing depends on a number of factors. One of the most essential requirements is good seed, which should have high vitality and good breeding. It should be free from disease and pests and suitable for the local conditions. When sown germination should be rapid giving healthy vigorous seedlings.ʌ

EXERCISE B *Adding statements*

Add one of the following statements for each of the caret marks ʌ in the paragraphs above.

(i) A clay loam may be suitable if there is no impermeable layer within three feet of the surface.
(ii) They should have an ample supply of manures and fertilizers.
(iii) Steam can be passed into mounds of nursery soil for about 24 hours.
(iv) The use of commercially prepared varieties is preferable to home grown stock.
(v) Chillis and egg plants should be planted in rows about 2 feet apart with a spacing of about 2 feet along the rows.
(vi) Vegetable crops can be produced in succession on the same plot.
(vii) Fruit requires a mixture with a higher level of potassium and phosphate in the later stages of growth.
(viii) Leafy vegetables are planted in raised beds about 6 to 9 inches apart.
(ix) Market gardening yields a much higher income than any other type of farming.
(x) Water can run off easily.
(xi) The seedlings will grow stronger and be better able to survive when planted out in the production beds.
(xii) A stiff hard clay should be avoided.
(xiii) Smaller varieties and those which mature earlier must be planted closer.
(xiv) A chemical sterilant such as methyl bromide can be used.

EXERCISE C *Relationships between statements*

Insert one of the following expressions in each of the statements you have just added to the paragraphs in Exercise A.

in this way	however
whereas	on the other hand
hence	therefore
in contrast	consequently
for example	for instance
alternatively	thus
for these reasons	generally

EXERCISE D *Contextual reference*

What do the following expressions refer to in the paragraphs you have written out?

(a) Paragraph 2: *varieties*
(b) Paragraph 5: *a mixture*
(c) Paragraph 5: *a higher proportion*
(d) Paragraph 6: *the soil*
(e) Paragraph 7: *varieties*
(f) Paragraph 7: *the aim*

II LANGUAGE IN USE

EXERCISE A *Directions and descriptions*

Statements which refer to the steps to be followed at various stages of growing particular crops may take the form of directions or descriptions. Study the following two columns:

GROWING SALAD PLANTS

I DIRECTIONS	II DESCRIPTION
1. *Nursery beds* Sow the seeds first in nursery beds. Disinfect the soil of the beds with boiling water. Mix the seeds with a little disinfected sand. Push the seeds into the soil, but not too deeply. Firm the soil well with the tamper. Water.	1. *Nursery beds* The seeds are first sown in nursery beds. The soil of these beds is disinfected with boiling water. The seeds are mixed with a little disinfected sand. They are pushed in the soil, but not too deeply. The soil is well firmed down with the tamper, and watered.

Part 1

Write out the following descriptions as sets of directions.

GROWING SALAD PLANTS

1. *Nursery beds*
See example.

2. *Transplanting*
The seedlings are lifted from the nursery beds about 3 weeks after sowing. They are transplanted at once. Rows are planted 30 cms apart with 30 cms between the seedlings. The earth is well packed down around each plant, and watered.

3. *Care of plants*
In hot climates a shelter is made over the salad plant beds. The ground between the plants is watered frequently. The weeds are removed by hoeing. Also, any snails and eelworms are removed. Diseased plants are taken out and burnt. Such plants are replaced with fresh seedlings from the nursery beds.

4. *Harvesting*
Salad plants are harvested when the leaves are crisp and green. They are picked in the morning or in the evening when the sun is not so hot. They are not picked when they are wet in case they rot during transport. Plants that have bolted are ploughed under after removing the seeds.

Part 2

Write descriptions based on the following sets of directions. Join sentences together where possible with *and*, *but* or *so*.

GROWING TOMATOES

1. *Tilling*
Prepare a deep soil which is friable and well drained.
Mix manure and fertilizers into the soil.
Apply potassic and phosphatic fertilizer.

2. *Nursery beds*
Sow seeds in nursery beds.
Disinfect the soil of the nursery beds with boiling water.
Sow the seeds in rows leaving 10 cms between the rows.
Do not push the seeds into the soil, merely cover them with a little earth.
Water twice a day.
When the seedlings begin to grow remove surplus seedlings and weeds.

3. *Transplanting*
Prepare the soil of the tomato beds several weeks before transplanting.
Work manure into the soil.
Put stakes in the beds.
Plant the seedlings in rows 50 cms apart with 40 cms between seedlings.
Transplant tomato seedlings 5 to 6 weeks after sowing.
Water the soil of the nursery beds, remove the seedlings without damaging the roots.

Transplant the seedlings at once.
Dig a hole at the base of each stake, lean the seedlings a little sideways, cover the roots with earth.
Pack the earth down well around each seedling, and water them.

4. *Care of plants*
Water the plants very often.
Cover the soil between the plants with cut herbage or leaves (mulching).
When cultivating remove weeds, snails and insects, take out and burn plants that are diseased or have been spoilt by insects.
Prune plants once or twice a month.
Keep only one or two main stems with their leaves and flowers.

EXERCISE B *Directions, descriptions and recommendations*

Recommendations tell you how something *should be done*. (See Unit 6, Language in Use, Exercise B).

EXAMPLE

GROWING SALAD PLANTS

1. *Nursery beds*
The seeds *should first be grown* in nursery beds.
The soil of these beds *should be disinfected* with boiling water and the seeds *should be mixed* with a little disinfected sand. etc.

Recommendations are often combined with descriptions and directions.

EXAMPLES

GROWING SALAD PLANTS

1. *Nursery beds*
The seeds *should first be grown* in nursery beds. The soil of these beds *is disinfected* with boiling water. *Mix* the seeds with a little disinfected sand. *Push* the seeds into the soil but not too deeply. The soil *is well firmed* with the tamper and *watered.*

OR The seeds *are first grown* in nursery beds. The soil of these beds *should be disinfected* with boiling water. The seeds *are mixed* with a little disinfected sand. *Push* them into the soil but not too deeply. The soil *should be well firmed* with the tamper and *watered.*

Write out paragraphs which give the procedures for growing tomatoes based on the information given below. Use directions, DIR, descriptions, DESCR, and recommendations, RECOM, as indicated, and produce two paragraphs ((a) and (b)) each time.
Add the following kinds of words where appropriate:

(a) definite and indefinite articles;
(b) pronouns such as *they* and *them*;
(c) forms of the verb *be*.

Join sentences together where possible with *and*, *but* or *so* and relative pronouns such as *which*. Compare your answers to those for Exercise A, Part 2.

EXAMPLE

GROWING TOMATOES

		(a)	(b)
1.	*Tilling*		
	Prepare deep soil, friable, well-drained.	DESCR	RECOM
	Mix manure, fertilizers into soil.	DIR	DESCR
	Apply potassic, phosphatic fertilizer.	RECOM	DIR

(a) *A* deep soil *is prepared which is* friable *and* well-drained. Mix manure and fertilizers into *the* soil. Potassic *and* phosphatic fertilizers *should be applied.*
(b) A deep soil *which is* friable *and* well-drained *should be prepared.* Manure and fertilizers *are mixed* into *the* soil. Apply potassic *and* phosphatic fertilizers.

GROWING TOMATOES

		(a)	(b)
2.	*Nursery beds*		
	Sow seeds in nursery beds.	RECOM	DESCR
	Disinfect soil of nursery beds.	RECOM	DESCR
	Sow seeds in rows leaving 10 cms between rows.	DIR	RECOM
	Do not push seeds into soil, merely cover with little earth.	DIR	RECOM
	Water twice daily.	DIR	DIR
	When seedlings begin to grow, remove surplus seedlings and weeds.	DESCR	DIR
3.	*Transplanting*		
	Prepare soil of tomato beds several weeks before transplanting.	RECOM	DESCR
	Work manure into soil.	RECOM	DESCR
	Put stakes in beds.	DESCR	DIR
	Plant seedlings in rows 50 cms apart with 40 cms between seedlings.	DESCR	DIR
	Transplant seedlings 5–6 weeks after sowing.	RECOM	DESCR

Water soil of nursery beds, remove seedlings without damaging roots. (use *can* in (a))	RECOM	DESCR
Transplant at once.	DIR	RECOM
Dig hole at base of stake. Lean seedlings sideways. Cover roots well with earth.	DIR	RECOM
Pack earth down well around each seedling and water.	DIR	RECOM

4. *Care of plants*

Water plants very often.	RECOM	DESCR
Cover soil between plants with cut herbage or leaves (mulching).	RECOM	DESCR
When cultivating, remove weeds, snails, insects.	DESCR	RECOM
Take out and burn plants diseased or spoilt by insects.	DESCR	RECOM
Prune plants once or twice monthly.	RECOM	DIR
Keep only one or two main stems with leaves and flowers.	DESCR	RECOM

EXERCISE C *Drawing conclusions*

Statements about requirements for crop growth can express conclusions deduced from general characteristics of the crop. Study the following information about *Salad plants.*

Characteristic	*Requirements*
grown for their green leaves	well prepared soil

We can express this information as follows:

Salad plants are grown for their green leaves.

So,	they *need/require* a well prepared soil.
Therefore,	the soil *should be* well prepared.
Consequently,	*it is important/essential that* the soil is well prepared.

Part 1

Organize the information in the following tables so that requirements match characteristics, and then use your tables to draw conclusions about requirements for the growing of tomatoes and salad plants, as in the example above.

TABLE 1

Characteristics	*Requirements*
1. Tomato roots go down deep into the soil	disinfect before sowing
2. Tomato plants grow very quickly	water and shade
3. Tomato seeds are easily attacked by disease	deep soil which does not retain water
4. Tomato fruits are very delicate	fine, rich soil
5. Young salad plants wilt in the hot sun	regular pruning
6. Salad roots do not have a large root system	careful handling at harvest time

TABLE 2

Requirements	*Recommendations*
1. Salad plants: fine rich soil	water every morning and evening and constant shelters over the bed
2. Salad plants: water and shade	grow at the beginning of the rotation
3. Tomato roots: deep soil which does not retain water	buy selected seeds or mix seeds with a powder
4. Tomato fruits: regular pruning to keep only one or two main stems	prune at least twice a month
5. Tomato seeds: disinfecting before sowing	till the soil fairly deeply
6. Tomato fruits: careful handling at harvest time	do not pile too many on top of each other

III GRAMMAR

Modal verbs

(i) The pattern *it + to be + possible, necessary, impossible, unnecessary,* etc.
(ii) The modal verbs *can, cannot; may, may not; must, must not; need, need not; should, should not; ought to, ought not to.*
If a sentence contains the pattern given in (i) above, *it is nearly always possible to rewrite it* using a modal verb from (ii).

= If a sentence contains the pattern in (i) above, *it can nearly always be rewritten* using a modal verb from (ii).

The following sentences show the relationship between the meanings of (i) and (ii). Study them carefully before doing the exercise. They are all concerned with rice cultivation.

(a) *It is possible to* give better care to seedlings in a nursery.	Better care *can* be given to seedlings in a nursery.
(b) *It is possible that* burning old rice stubble will reduce diseases in some instances.	Burning old rice stubble *may* (or *can*) reduce diseases in some instances.
(c) Inter-row weeding *is impossible to* do when seeds are sown broadcast.	Inter-row weeding *cannot* be done when seeds are sown broadcast.
(d) *It is possible that* the field is *not* even and so some plants will be deeper than others.	The field *may not* be even and so some plants will be deeper than others.
(e) The use of mechanical weeders *is permitted* when the plants are grown in a straight line.	Mechanical weeders *may* (or *can*) be used when the plants are grown in a straight line.
(f) With the broadcast method *it is essential not* to flood the fields deeply until the seed roots have anchored in the ground.	With the broadcast method the fields *may not* (or *must not*) be flooded until the seed roots have anchored in the ground.
(g) With broadcast sowing *it is essential that* the levelling of the bed is accurate.	With broadcast sowing, the levelling of the bed *must be* accurate.
(h) *It is not essential to* grow rice seedlings in nurseries.	Rice seedlings *need not* be grown in nurseries.
(i) *It is advisable to* soak pre-germinated seed for 24 hours before use.	Pre-germinated seed *should be* (or *ought to be*) soaked for 24 hours before use.
(j) *It is not necessary* to grow some varieties of rice in ponds.	Some varieties of rice *need not* be grown in ponds.
(k) It is *essential not* to use non-selective weed killers for rice plants.	Non-selective weed killers *must* (or *may*) *not* be used for rice plants.

(l) *It is inadvisable* to build nursery beds more than five feet wide.	Nursery beds *should not be* (or *ought not to be*) more than five feet wide.
(m) *It is necessary* to grow wet paddy or swamp rice in water.	Wet paddy or swamp rice *must be* grown in water.

EXERCISE

Rewrite the following sentences using one of the verbs given in brackets instead of the expression in italics. Make any other changes in the sentence that are necessary.

1. *It is necessary that* the fields of plots used for rice growing are surrounded by bunds which hold the water in the field. (must, can, may)
2. An advantage of growing rice seedlings in nurseries is that *it is possible to* adjust the transplanting time to a period when there is sufficient rain for flooding. (can, should, will)
3. *It is* always *necessary for* the nursery to be in the open sunlight and on the most fertile and level soil. *It is desirable, too, that* water is available for the nursery. *It is essential that* the beds are surrounded by a low wall and are not more than 5 feet wide so that *it is possible to* look after the beds easily. (should, can, must; may, will, should; must, may, will; must, can, will)
4. As the seedlings grow in nursery beds increasing the depth of the water *is possible*, but more than about 1 inch is *never advisable* and, *as a rule*, the upper parts of the seedlings are never covered. (may, must, needn't; needn't, should, will; can, needn't, must)
5. Rice grows best on clayey loams that become muddy when puddled, but cultivation on damp alluvial soils, light sands and gravelly and stony soils *is also possible*. (can't, can, must)
6. To hold irrigation water evenly *it is essential* to level paddy fields; therefore, on hill slopes *it is necessary that* the land is terraced. (may, must; should, needn't, must)
7. During or just before transplanting nitrogen fertilizer *is certain to* help the newly planted shoots. (should, will, may)

For the rest of the exercise, choose the most appropriate modal verb for yourself.

8. During winnowing *it is important that* rice grains are clean and not mixed up with earth and stones.
9. In puddle cultivation *it is possible to* germinate the seed before sowing by soaking it in water.
10. Before sowing, *it is recommended that* the seed is treated with Agrosan GN (fungicide).
11. *It is best to* keep nursery crops free of weeds; watering by hand and protection against insect pests and diseases are *both advisable* too.

12. Rice responds well to nitrogen application in organic or inorganic form, but *it is possible to* add superphosphates at 20–40lb of P O per acre during puddling.
13. *It is necessary to* use chemical weed killers MCPA and 2:4-D to combat weeds with the broadcast method of sowing.
14. If planting is deeper than 2 inches in depth, growth *is certain to* be retarded, and, if less, *it is possible that* the seedlings will lift out.

IV SUMMARIZING

Summarize the following instructions in note form. Draw up tables with headings for each column as follows:

1. Soils and fertilizers.
2. Planting and spacing.
3. Cultivation and diseases.
4. Harvesting.

1. GROWING BEANS

To grow beans the soil must be well prepared. The soil is tilled deeply so it will hold moisture, and the roots can grow down deeply. Beans grow in light soils or in soils rich in humus.

Potassic or phosphatic fertilizers should be used, or a compound fertilizer such as 'Nitrofoska red'.

It is best to grow beans at the end of the rotation. In that way they use up the mineral salts still left in the soil. It is essential to buy new seeds every year.

Sow directly into open beds. Put three seeds in each hole 3 to 4 cms deep. For pole varieties you should leave 1 m between the rows and 70–125 cms between the seed holes. A trellis of sticks 2 m high must be constructed.

When the seedlings come up the weeds should be removed. Heap up the soil around the base of the plants but be careful not to damage the roots.

The main insects are yellow spiders and aphids. Treat the leaves with Phosdrin. Rust can be prevented by treating with 1% Bordeaux mixture.

The pods should be picked every day, otherwise no new pods will form and the harvest will be less plentiful. When picking the ripe pods be careful not to damage the young pods which are forming.

2. GROWING ONIONS

The soil must not be too moist. It is essential, therefore, to till deeply so that the water goes down and air can get in. The soil must be rich in humus.

Apply fertilizers. Onions need above all potassium and phosphorous. Be careful not to apply too much nitrogen, or the leaves will develop more than the bulb.

It is best to grow onions after salad plants. Never grow two crops of onions after each other because of the possibility of disease. Never sow seeds which are more than 1 year old.

Small fields: it is better to sow the onions in nursery beds and to transplant later when the seedlings have grown 15–20 cms. Leave 20 to 30 cms between the rows and, along the row, 10 to 15 cms between the plants. Pack down the earth well around each plant.

Large fields: sow in open beds, 25 to 30 cms between plants.

When sowing in nursery beds the soil should be disinfected. It should have a fine tilth and be firm in depth. Make a shelter over the beds and water the beds twice a day. Surplus seedlings should be thinned out and weeds removed.

Eelworms attack the roots and the base of the onion. Infected plants should be pulled up and burnt and the soil disinfected with boiling water or with Vapam or Nemagon. When downy mildew appears, treat the plants with copper based products.

Onions should be harvested only when they are quite ripe. It is best to lift the onions when it is not raining so they will not get wet and rot. Leave them lying on the fields for a few days to dry. Cover with a little grass or straw. To keep onions well, they are stored in a dry, airy place which is well protected against rats or other animals.

8 Animal Husbandry

PASSAGES FOR COMPREHENSION

PASSAGE I

● *The digestive system of farm animals*

EXERCISE A

Read the following passage and write in your notebook the appropriate expression to fill the blank spaces, (a) to (g). You should choose from the following expressions:

for example, therefore, however.

When an animal eats, the food passes along the digestive tract. This is the tube which runs from the mouth to the anus, or, in the case of poultry, the vent. It is known as the alimentary canal. Into this tube digestive juices, or enzymes are secreted. These enzymes break down the food into a form which the animal can digest, or absorb into its body through the bloodstream and use. The undigested part is rejected in the form of excrement. (a) . . . . . ., food protein is broken down into amino acids. These substances are the basic building blocks of body protein. They are absorbed into the body and used to build new proteins in the animal such as muscle and milk.

Different types of farm animals have different alimentary tracts. Cattle, (b) . . . . . ., have highly developed stomach systems which are made up of four parts. (Fig. 1) When a cow eats grass, it does not chew it but swallows it at once. The grass passes down the oesophagus to the first stomach, or rumen, where it is stored. When the cow has filled its first stomach, it often lies down. But, it goes on moving its jaws. This is because it is ruminating. The cow brings up, or regurgitates, some of the grass from its first stomach for a second chewing, or cudding, with its molar teeth. When the grass is well chewed and broken down, the cow swallows it again. This time, (c) . . . . . ., the food goes into the second stomach, or reticulum, where organisms start to attack the cellulose in the grass in order to release the nutrients which are enclosed with-

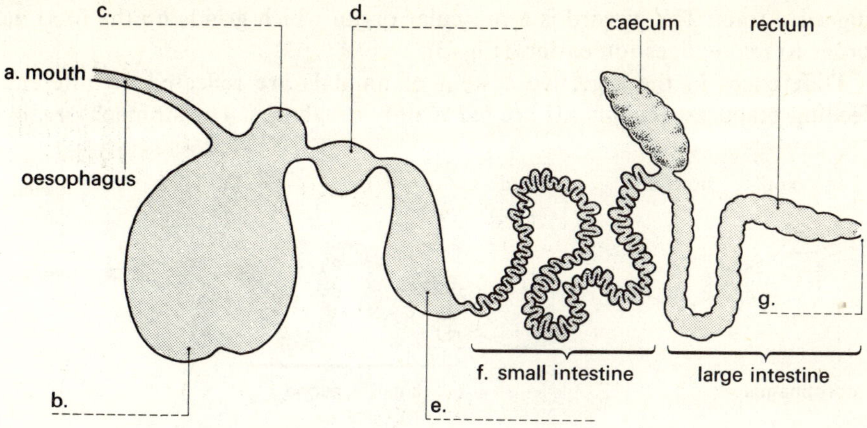

FIG. 1 *Alimentary tract of a cow*

in the cells walls. From here, the food passes through the omasum into the fourth stomach, or abomasum, into which the digestive juices are secreted. These enzymes kill the organisms in the food and make them available for digestion. Digestion and absorption of nutrients takes place in the small intestine. What remains undigested is then rejected through the anus as excrement.

Cows and other animals with similar stomach systems ruminate because their digestive systems are made to digest and transform roughages. These are fibrous materials such as hay, grass and straw. Hence, cows, goats, sheep and camels are known as ruminants.

Non-ruminants such as pigs and poultry have a single or simple stomach in which the acid digestive juices exist. (Fig. 2) Such stomachs, (d) . . . . . ., are

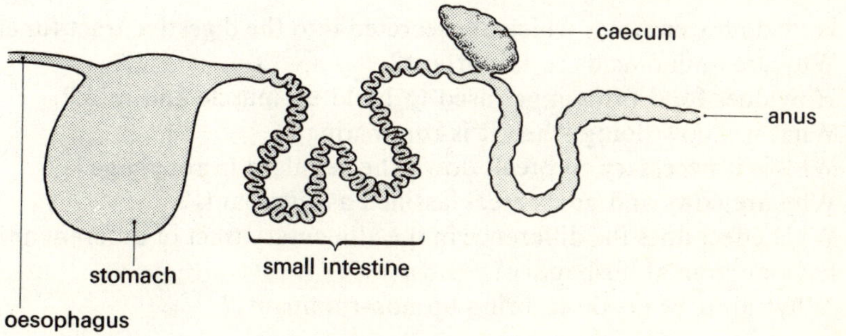

FIG. 2 *Alimentary tract of a non-ruminant*

unsuitable for the activity of organisms. In these animals the caecum is the place where there is most organism activity. (e) . . . . . ., since the products of this organ will pass out of the animal, the activity of organisms in the caecum is of limited value. Poultry have a crop which stores food before it enters the digestive tract. The gizzard is a muscular organ which grinds up the food in order to make digestion easier. (Fig. 3)

Differences in the digestive system of animals are reflected in different feeding practices. Ruminants are fed mainly roughages. The stomach organ-

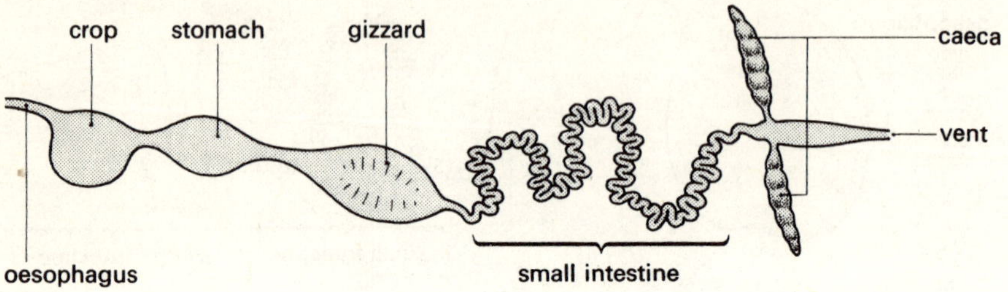

FIG. 3 *Alimentary tract of poultry*

isms use the protein in the roughage and non-protein nitrogenous substances to manufacture their own body protein. Non-ruminants can make little use of hay and straw, unless the caecum is highly developed, as it is in the horse. This is because no conversion of the roughage takes place in their stomachs. (f) ......, non-ruminants are fed on higher quality proteins such as barley and maize or fish and bone-meal. (g) ......, the very young ruminant is treated as a non-ruminant as regards food protein until its stomach is fully developed.

EXERCISE B

Answer these questions:

(a) How do the enzymes which are secreted into the digestive tract function?
(b) Why are amino acids so important?
(c) How does food protein get used to build up muscle and milk?
(d) What is a cow doing when it is ruminating?
(e) Why is it necessary to break down the cellulose in roughage?
(f) Why are cows and goats etc. classified as ruminants?
(g) What effect does the difference in the alimentary tract of different animals have on animal husbandry?
(h) Why can it be costly to bring up non-ruminants?

EXERCISE C

Draw a diagram like the one on the next page representing the digestive tract of a cow. By referring to the passage and Figure 1, label the different parts of the system, (a) to (g), and write appropriate labels to describe the separate stages in the digestive process, 1–8. Two labels, (a) and (f), and two descriptions, (1) and (7), have been written for you.

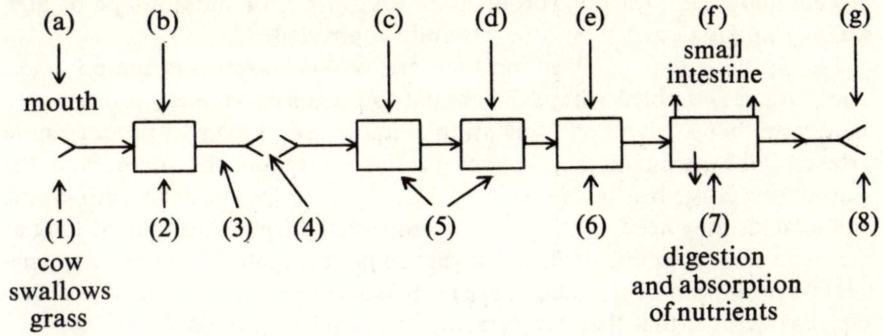

PASSAGE II

● *Food requirements*

EXERCISE A

Read the following passage and write in your notebook the appropriate expressions to fill the blank spaces, (a) to (g). You should choose from the following expressions:

for example, however, on the other hand, therefore.

Food is made up of varying proportions of several substances. These substances include water, carbohydrates, fats and oils, proteins and minerals.

Water has a number of functions. It carries nutrients from the intestinal wall to the cells and tissues. This process of transportation is carried out by the blood, the basis of which is water. Another function is to facilitate digestion by assisting the passage of food through the alimentary tract.

Carbohydrates consist of sugars and starches which are broken down into glucose during digestion. They provide the animal with the energy needed to perform the normal functions of life and movement. These include, (a) ......, the production of milk in lactating animals, and the ability to work in draught animals. Fat and oil are also a source of energy. They are changed to liquid form in the reticulum, where organisms reduce some of the fat to glycerol and fatty acids. (b) ......, as fat and oil are valuable for human consumption, the fat content of most foodstuffs used for animal feeding is low. This is because the oil is first extracted from the oil seeds and only the remaining residues are used for animal feed.

Fibre consists mainly of cellulose and lignin. Its nutritive value varies according to species. Ruminants, (c) ......, can easily digest cellulose, whereas animals with simple stomachs are able to use only small quantities of fibre efficiently. Fibre gets broken down by the rumen bacteria into substances which can be used as a source of energy. In the case of a lactating cow a low intake of fibre may result in milk with a low percentage of fat. (d) ......, a high proportion of fibre in the diet prevents the consumption of sufficient protein and starch for milk production. (e) ......, the fibre intake of high yielding milking cows should be carefully controlled.

The proteins that animals consume are broken down into amino acids. These are reassembled within the animal to form its own body protein such as muscle, bone, blood etc. Lactating animals use protein to produce milk proteins; the pregnant animal uses protein to enable the foetus and the placenta to grow. It is not necessary, (f) ......, to feed animals with *all* the amino acids they need. Some can be synthesized during digestion, as long as there are enough amino acids in the diet to permit synthesis. Those that *cannot* be synthesized by the animal are termed *essential* amino acids. It is necessary, (g) ......, that they are present in the dietary protein.

Minerals play a vital part in the structure of the animal's body and its functions. The most important minerals for dairy cows are calcium, phosphorous and chlorine. A deficiency of calcium can give rise to milk fever in lactating animals. Phosphorous deficiency may cause sterility. To ensure a satisfactory supply of these major minerals the following mineral mixture can be added to the diet of dairy cattle at $2\frac{1}{2}$ lb per cwt of concentrates:

Common salt	1 part by weight
Sterilized steamed bone flour	1 part by weight
Ground limestone	$\frac{1}{2}$ part by weight

For animals grazing during the time the pastures are growing rapidly, as in the spring and autumn, a magnesium supplement may be necessary. Magnesium deficiency in this rapidly growing herbage may cause the disease known as *hypomagnesaemia* or *grass staggers*. Animals which are affected by this disease show nervous symptoms such as muscular spasms, sweating and grinding of the teeth. To prevent this, a supplement of 2oz of calcined magnesite per cow per day should be fed to the animals.

EXERCISE B

Answer the following questions.

(a) Give two reasons why water is essential to livestock.
(b) In what way are carbohydrates essential for lactating animals?
(c) Why are only the residues of oil seeds fed to animals?
(d) Why are non-ruminants not fed large quantities of fibrous food?
(e) Give two reasons why care should be taken feeding fibrous food to lactating cows.
(f) What part do amino acids play in the dietary system of livestock?
(g) How do essential amino acids differ from non-essential amino acids?
(h) What may be the result of insufficient calcium in the diet of a milking cow?

(i) How can we make sure that the diet of farm animals is not deficient in major minerals?
(j) What effect on the health of an animal may rapidly growing pasture have, and how can it be avoided?

EXERCISE C

Copy the following tables into your notebook and complete them with information from the passage.

(a)

SUBSTANCE	FUNCTION	OPERATION
water	(a) carries nutrients from intestinal wall to cells and tissues (b)	(a) transported by the blood (b) ......
......	provide animal with energy needed to perform normal functions of life	......
fats and oils	......	changed to liquid form and ......
fibre	......	......
......	form body protein of animal such as muscle, bone, blood, etc.	......

(b)

DEFICIENCY	DISEASE OR CONDITION	REMEDY
calcium	......	
......	sterility	......
......	......	Add 2 oz of calcined magnesite.

PASSAGE III

● *Feedingstuffs*

Feedingstuffs used on farms can be divided into two main groups: bulk foods and concentrates. Bulk foods contain a limited amount of nutrients in a given weight of food. Hay contains roughly 35 lb of starch per 100 lb weight. Concentrates are richer in nutrients. Barley contains twice as much starch per 100 lb weight, or twice the amount of energy in an equivalent weighing of foodstuff. Both these groups of foodstuff are used in feeding livestock. The bulk foods satisfy the animal's appetite. Concentrates are given to ensure that the animals have taken in sufficient nutrients. Ruminants are particularly suited to consume bulk foods, because of their ability to break down fibre, but their diet should be supplemented with concentrates. The rations of non-ruminants consist mainly of concentrated foods because their simple stomachs cannot utilize fibre.

Bulk foods consist of coarse fodders such as hays and straws, and succulent foods such as roots, green fodders (kales and grass) and silage. The former are low in protein and minerals; the latter have a high moisture content which makes them palatable. Three types of hay can be distinguished: meadow hay, which usually contains a large number of grass species and clovers; clover hay, which is produced from fields planted with a clover crop; and seed hay which is more fibrous and coarse than meadow hay. Because hay is such a commonly used food a table of 'hay equivalent' has been drawn up. This table can be used to calculate the amount of substitute food which would be required to replace part of the hay in a food ration.

TABLE 1 *Hay equivalents*

1 lb of average hay	=	½ lb of barley
	=	2 lb of oat straw
	=	3 lb of silage
	=	4 lb of kale or cabbage
	=	5 lb of mangolds or swedes

Concentrates can be sub-divided into foods rich in carbohydrates such as oats, barley, wheat etc.; protein rich foods of vegetable origin, such as oil cakes and meals of linseed, groundnuts or soya beans; protein rich foods of animal origin including fish meals or meat meals; and home-grown foods rich in protein including leguminous crops such as peas and beans. The production of high quality home-grown feed is very important to reduce costs.

Balance is an essential element of animal feeding. The balance of the total food ration should be correct otherwise the animal will be receiving an excess or deficiency of starch or protein. An imbalance in feeding ration may lead to unnecessary expense and so reduce profits. Balance also applies to the effect which the foods in the total diet have on the digestive system. The best balance is a feeding ration which is palatable and mildly laxative. In addition, minerals and vitamins must be balanced to keep the animal healthy and productive. As we have already noted, an incorrect balance in the calcium-phosphorous ratio can lead to infertility in cattle.

EXERCISE A

Answer the following questions.

(a) On what basis can feeding stuffs be classified?
(b) If 100 lb of hay and barley are compared, what percentage of each consists of starch equivalent?
(c) Why does the diet of ruminants usually consist of both bulk foods and concentrates?
(d) What characteristic of foods such as green fodder and silage make them valuable in a diet?
(e) What does a table of hay equivalents show?
(f) How can a livestock farmer's costs be reduced?
(g) What effect can an imbalance in food rations have (i) on the animal, and (ii) on the farmer?
(h) What harm can a mineral imbalance in the diet have on an animal?

EXERCISE B

Classify feedingstuffs in the form of a diagram as in Unit 4, Language in Use, Exercise C.

EXERCISE C

Refer to the table of *Hay equivalents* in the passage and complete the paragraph below.

To provide maintenance for an 11 cwt cow, 10 lb of hay could be replaced by ...... of grass silage. This gives a feeding ration, in place of 20 lb of hay, of 10 lb of ...... and 30 lb of ...... Alternatively, if hay were in short supply, 2½ lb of ...... and 5 lb of hay could be given, which is the equivalent for ...... of hay, and, to make up the 20 lb hay equivalent feeding ration, 20 lb of ......

PASSAGE IV

● *Reproduction*

EXERCISE A

The blank spaces in the second paragraph of the following passage can be filled with expressions from Figure 1. Rewrite the paragraph, inserting the correct expressions. Similarly, the labelling of Figure 2 can be completed using expressions from the third paragraph. Draw the diagram and complete its labelling.

The mating of animals, or reproduction, is extremely important in animal husbandry. It is the farmer's duty to make sure that mating is carried out correctly and that the animal is well looked after during its pregnancy and at the time of birth.

The female reproductive organs consist of two ...... from which the eggs, or ova, originate, and two ...... or oviducts down which the eggs travel after being released. The ...... are connected to the ......, or womb, where the egg is implanted and the young animal develops. The cylindrical passage in which the male spermatoza is deposited is known as the ...... The external opening of the reproductive tract is the ...... Between the ...... and the body of the ...... is a narrow constriction called the ......, which can be closed to prevent infection entering.

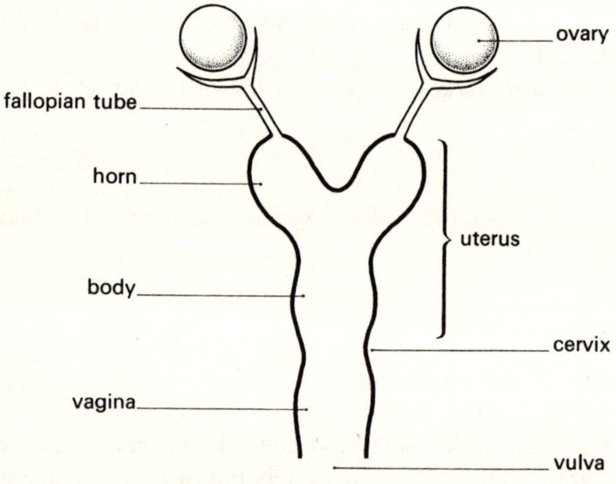

FIG. 1 *Reproductive organs of the cow*

The male reproductive organs consist of two testicles, or testes, where the spermatozoa are formed. The sperm travels down the vas deferens. There are three sets of glands which secrete the fluid substance which forms the semen in which the sperm swim. These are the seminal vesicles close to the bladder, and just below it the prostrate gland, and below that the bulbo-urethral gland. The urethra is a common passage for both sperm and urine in the male. It ends in the penis.

At the time of ovulation the egg or eggs which are released pass into the fallopian tube, together with the fluids from the follicle. The egg remains at the top of the tube for a while in a clot of fluids. When this clot dissolves the egg passes down the tube into the uterus. While the egg is in this clot, or just

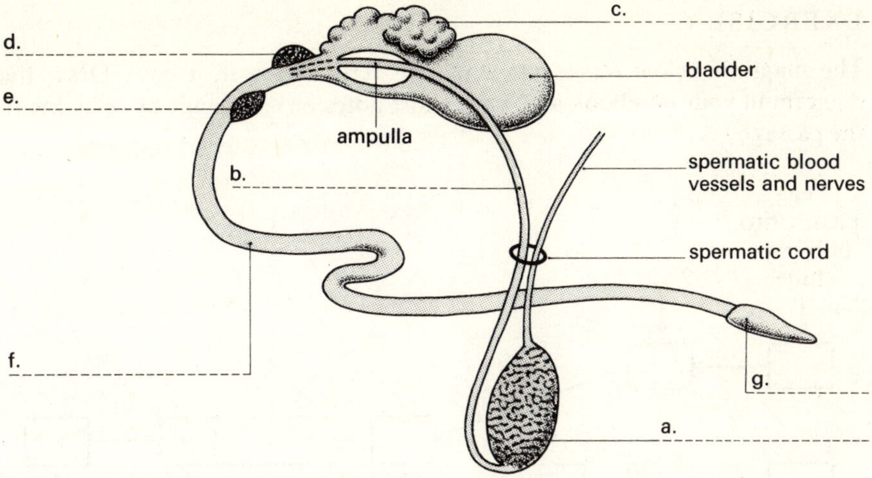

FIG. 2 *Reproductive organs of the bull*

emerging from it, fertilization – that is, the union of the sperm and the egg – occurs. At the time of mating the male sperm is deposited in the female tract near the cervix. It therefore has to travel the full length of the uterus and up the fallopian tube to fertilize the egg. Only one sperm enters the egg although several may swim around it. After that the two nuclei; one from the sperm and one from the egg, are joined, or fused. The young animal has now been conceived. The egg moves down the fallopian tube into the uterus where it is implanted. If it has been fertilized, it begins to divide. The egg continues to divide on its journey down the fallopian tube. In cows and sheep it reaches the uterus between seventy and ninety hours after ovulation.

EXERCISE B

Which of the following statements about the reproductive organs of cattle are true or false?

(a) The ova originate in the fallopian tube.
(b) The ova travel down the oviducts after their release.
(c) The male spermatozoa is deposited in the horn of the uterus.
(d) The cervix is the constriction at the base of the womb.
(e) The seminal vesicles are below the bulbo-urethral gland.
(f) The spermatozoa swim down the vas deference in semen secreted by the testes.
(g) The urethra is the tube running from the bladder to the penis.

Rewrite correctly those statements which are false.

EXERCISE C

The diagram below represents stages in fertilization in a cow. Draw the diagram in your notebook and write short notes on each stage by referring to the passage.

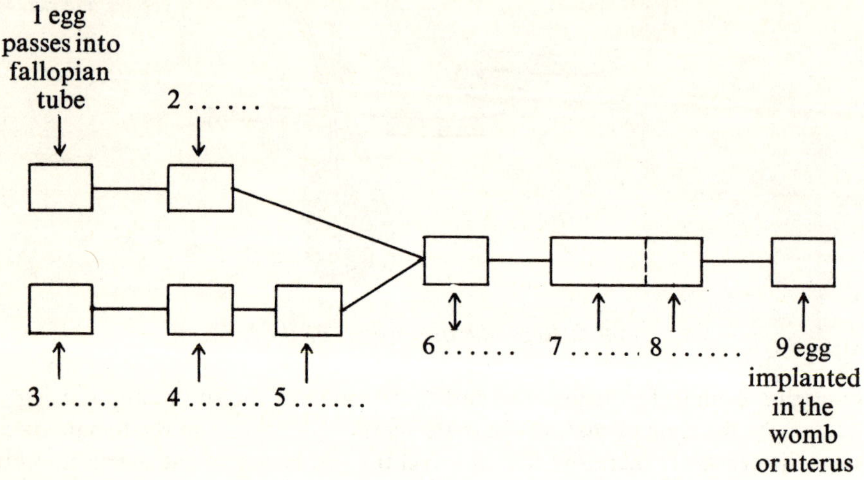

PASSAGE V

● *Breeding*

EXERCISE A

Read the following passage and write in your notebook the appropriate expression to fill the blank spaces, (a) to (i). You should choose from the following expressions:

for example, therefore, however.

If an Aberdeen Angus bull is mated to an Aberdeen Angus cow the resultant calf will have the black coat and stocky appearance of its parents. It will also be without horns, or polled. Coat colour and the presence or absence of horns are two examples of inherited characters. If this calf is mated with an animal of the same breed its calf will show the same inherited characteristics. Every aspect of the animal is controlled by the characters it has inherited from its parents. (a) ......, the breeder aims to produce an animal which has inherited the best possible characters.

These characters are chemical substances known as genes. In genetics, the letters of the alphabet are used to indicate characters or genes. The gene for red colour, (b) ......, may be written R, and for white W. A red bull will have a pair of genes in each nucleus RR. Similarly with a red cow. If these two are mated the calf will obtain R from its sire and R from its dam. (c) . . . . . ., its genetic make-up will be RR. In matings of cows and bulls with identical genes each animal is said to be *homozygous* for that particular colour. The red bull had RR and was, (d) . . . . . ., homozygous for red. If, (e) . . . . . ., the red bull is mated to a white cow, having WW genes, the calf will receive R from its sire and W from its dam so that its genetic make-up will be RW and its colour between red and white, or roan. This calf will not possess identical genes and it said to be *heterozygous* for colour.

In some cases a gene for a particular character may not allow the effects of the other gene to show. In other words, it suppresses the expression of the other gene. Such a gene is then said to be *dominant* to the other gene; and the suppressed gene is said to be *recessive*. In cattle, (f) . . . . . ., black coat colour is dominant to red, and the polled gene dominant to the horned gene. A polled bull which is homozygous for the polled character will be PP. A horned cow will be **pp**, denoting the recessive gene. The bull will produce sperm with one gene P; the eggs of the cow similarly will contain one gene **p**. When these two animals are mated there is only one possible way in which the genes can be arranged in the fertilized egg, namely **Pp**. (g) . . . . . ., all the calves from the mating will be **Pp** and will be heterozygous (Fig. 1).

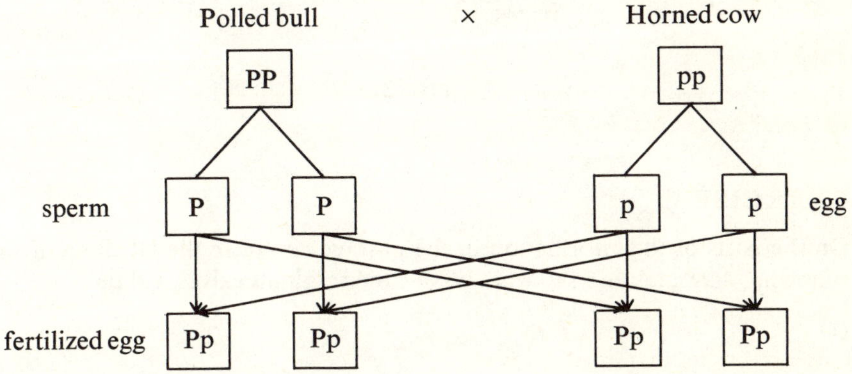

FIG. 1

When heterozygotes are mated the homozygous recessive gene will occur on average once every four calves. (h) . . . . . ., a bull which is heterozygous to coat colour will appear black because black is dominant to red. When mated to a similar heterozygous cow, on average two heterozygous calves and one homozygous calf will be produced which are black in colour. Occasionally, (i) . . . . . ., one calf will be red, a homozygous recessive, in other words. This is why one sometimes sees red Aberdeen Angus calves. (See Fig. 2 below).

EXERCISE B

Draw the following diagram in your notebook and complete the labelling.

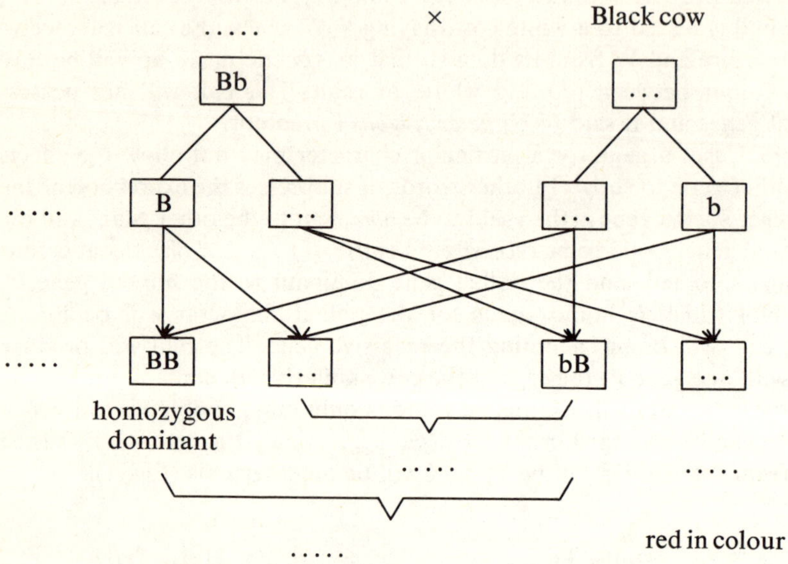

FIG. 2

EXERCISE C

On the basis of the information in the passage complete the labelling of the following diagrams and say what colour the resultant calves will be.

(a)

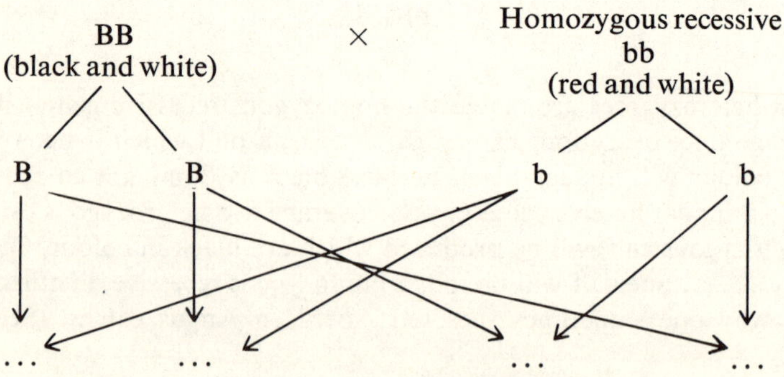

(b)

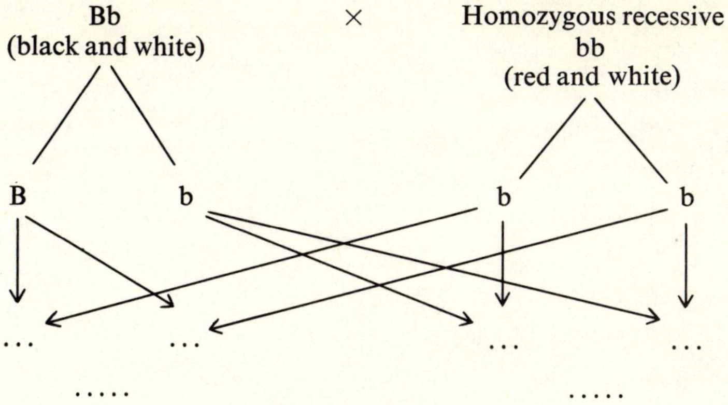
Bb
(black and white)
×
Homozygous recessive
bb
(red and white)
B
b
b
b
...
...
...
...
.....
.....